PREFACE TO FIRST EDITION

INTERFEROMETRY is an elegant branch of optics which has made considerable contributions in a variety of different fields in physics.   It is elegant primarily because of its economy of means. It is indeed striking that with little more than two slits or two mirrors one can, on the one hand, plumb the depth of space to measure the diameter of a star or, on the other hand, the heart of a crystal to measure the sizes of molecules.   The essential success with interferometry lies not so much in the means at disposal but in the critically correct use of these simple means.   The subject therefore constitutes an admirable exercise in discretionary application and as such has much pedagogic value.

Basically interferometry has acquired importance by making available versatile and sensitive tools for a wide variety of physical measurements, and especially characteristic is the extensive range of interests to which it serves as a handmaiden.   Thus, on the one hand, interferometers yield values for the magnetic moments of atomic nuclei, and, on the other hand, they can give the magnetic moment of the sun.   It is self-evident that so ubiquitous a technique must occupy a significant place in a degree course in physics.

There exists a very real need for an up-to-date elementary introductory text on interferometry, and it is the absence of such a text for my own students that has induced me to prepare this one.   The insignificantly small number of books published on the subject during the past fifty years have all rather tended to be somewhat specialized, concentrating, perhaps to advantage, on the specialized fields of interest of the authors, i.e. they have been of the nature of monographs rather than general elementary introductions.   This text is essentially an elementary introduction for degree students.

I am very conscious of the fact that an undergraduate to-day has but limited time to devote to interferometry.   Optics may perhaps occupy one-fifth of his physics course, and interferometry may possibly constitute but a fifth of the optics to be learnt. Indeed a student may find at his disposal perhaps no more than two hours a week to devote to this aspect.   I am quite convinced that under such conditions it is useless to expect him to refer to original papers, many long, many in a foreign language.   I have therefore deliberately left out all references as being in the circumstances superfluous.   However, there may be here and there an odd enthusiast who might wish to pursue the subject further. For him I include at the end a brief bibliography of some specialized books on interferometry published during the past fifty years. From these he will find ample references to keep him reading many a year.

A brief comment as to the contents.   Publication in the field is vast and some thousands of papers can be listed, thus a rigid selection is essential.   In the main my approach has been historical.   The book is based (with appreciable additions) upon the lecture course which I give to my second-year students who are reading for Part I of the B.Sc. Special Degree in Physics at London University.   It should cover any typical degree course in physics.   In such a course I have always found a logical difficulty about where to introduce the Michelson echelon and the Lummer plate, feeling that they should be preceded by the much more important Fabry-Perot interferometer.   I have therefore relegated these two instruments to a brief concluding chapter. Chapters 1 to 12 and Chapter 15 are the important ones for a degree course, the rest should be treated as optional since they have so strong a personal bias.

With the object of indicating to the student that interferometry is still very much a live subject, I include as illustrations of current work a number of photographs taken in my own laboratory over the past ten years.   I am glad to take this opportunity of thanking my technician Mrs V. Hinton for assistance in the preparation of the diagrams and illustrations.

I wish to express my thanks to Methuen & Co. Ltd. for permission to extract some five pages from my book *High Resolution Spectroscopy* (1947).   I also wish to express my thanks to the Clarendon Press, Oxford, for permission to extract some five pages from my book *Multiple-beam Interferometry* (1948).

October, 1954

An Introduction to
INTERFEROMETRY

By the same author

INTRODUCTION TO ATOMIC PHYSICS

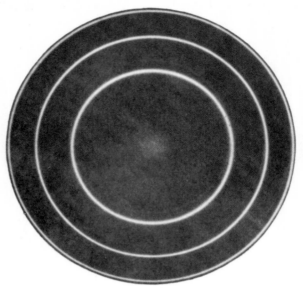

(a) Fabry-Perot fringes given by a monochromatic line source.
The fringe sharpness should be noted.

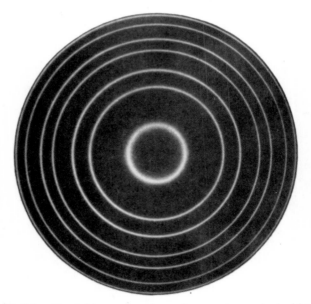

(b) Fabry-Perot fringes given by a close spectroscopic doublet.
Resolving power is high and the two components are so well
separated that the one lies nearly (but not quite) half-way
between the orders of the other.

An Introduction to

INTERFEROMETRY

SECOND EDITION

S. TOLANSKY
Ph.D., D.Sc., F.R.S.

Professor of Physics
Royal Holloway College
London University

LONGMAN

PHYSICS

Longman Group Limited
London

*Associated companies, branches and representatives
throughout the world*

© Longman Group Limited 1955, 1973

First published 1955
Second Impression by Photolithography 1960
Third Impression by Photolithography 1962
Fourth Impression by Photolithography 1966
Second Edition 1973

ISBN 0 582 44333 4

Printed in Great Britain by
William Clowes & Sons, Limited
London, Beccles and Colchester

PREFACE TO SECOND EDITION

SINCE writing the previous edition (1954) there have been two notable extensions in the applications of optical interferometry. The one is the development and use of many different kinds of interference microscope, now manufactured on an international scale. Instruments have been developed separately, mainly for the metals scientist and for the biologist. A new chapter has been added dealing with these developments. The second extension of interferometry, following the invention of the laser light source in 1961, has been the phenomenal outpouring of publications in that branch of applied interferometry called holography. An introductory section on the laser is therefore included, followed by a chapter which is a summary of the principles, practice and some typical recent applications of holography. This subject is expanding explosively.

In addition to these two new chapters, opportunity has been taken to remove some infelicities in the former edition, some diagrams have been improved and some derivations of formulae simplified. Some extra material has been added to the section on polarized light, and to the section on the evaluation of the metre.

The text, as before, is strictly restricted to what has long been accepted as optical interferometry. Hence the current extensive applications of interferometry both to acoustics and to radio (including radio astronomy) are excluded in the present text.

The level aimed at here is for the second year of an honours degree course in physics. The student is so overloaded with modern physics that classical subjects, like interferometry, must needs be pruned and compacted. This has been a guiding principle in the preparation of this text. The bibliography has been slightly expanded.

S.T.

CONTENTS

LIST OF PLATES

CHAPTER 1

THE NATURE OF INTERFERENCE

Historical

The earliest studies of interference in optics are intimately bound up with the evolution of the theories of light propagation and ultimately settled a long-standing conflict between the wave and the corpuscular theories of light.   Interference effects in optics were known at an early date.   In 1665, Grimaldi, a careful and accurate observer, discovered the diffraction fringes produced both by a narrow obstacle and by a slit.   He tried to explain these observations by supposing light to consist of a fine fluid in a state of vibration.

In the same year Hooke attempted to explain the colours of thin films (discovered two years earlier by Boyle) also by means of a crude wave theory.

Huygens, in 1678, then announced his principle of wave propagation through subsidiary wavelets and a few years later attempted to account for the double refraction of calcite which had been discovered eight years earlier by Bartolinus.   This he failed to do because the conception of transverse vibrations had not yet emerged and the longitudinal vibrations he postulated could not explain the phenomena of polarization.

During this period (since 1666 in fact) Newton had been conducting his researches in light, culminating in the publication of his *Opticks* in 1704.   The first part of this epoch-making work dealt exclusively with refraction, dispersion and the discovery of the spectrum.   The second part was devoted to the study of the interference colours of thin films and especially to the phenomena now called Newton's rings.   The third and last part covered a reinvestigation of Grimaldi's experiments on diffraction.   Newton discussed both a wave and a corpuscular theory of light propagation.   He was unable to reconcile rectilinear propagation with a wave theory and pointed out that the double refraction of calcite implied a polarization, which was inadmissible with longitudinal vibrations, *the only type then recognized*.   He therefore developed

1

his celebrated corpuscular theory, introducing the hypothesis of "fits of easy transmission and reflection". By this means he accounted for the Newton's rings and indeed was able to calculate with surprising accuracy the quantity we now call the wavelength. He explained diffraction as being due to an "inflexion" of the light corpuscles passing close to an edge.

The whole of the eighteenth century was domineered by the impressive authority of Newton, and no further progress was made until Thomas Young, in 1802, proved by simple but brilliant experiments that light is subject to interference and is propagated as a wave form. Young was the first to realize both the Principle of Coherence and the Principle of Superposition, both of which are fundamental in interference. His work was followed in 1815 by those outstanding developments associated with Fresnel, who gave the death blow to those doubters and critics who still obstinately refused to be convinced by the decisive experiments of Young. Fresnel also laid the firm foundation of a theory of light propagation and light diffraction in a manner still valid.

Other discoveries were rapidly made in the field of interference. Herschel, in 1809, discovered highly sharpened fringes in a thin film when light emerges near to grazing incidence (Herschel's fringes). These remained an obscure curiosity and almost a century passed before they were fully explained. Eight years later (1817) Brewster reported the interference effect (Brewster fringes) which takes place when light passes through or is reflected by two similar plane parallel plates slightly inclined to each other. Lloyd performed an important interference experiment using a single mirror in 1837, and in 1849 Haidinger observed and accounted for the rings now known by his name, using a piece of mica to produce the phenomenon. From the middle of the nineteenth century onwards many new discoveries were made especially by Fizeau. New forms of interference were studied and many of these were applied to widely different fields. A considerable body of publications grew up until now the subject of "Interferometry" is one of much complexity and importance. It has by no means ceased to be active. New interference procedures are still reported, but of much greater significance are the intensive applications of interference methods to many fundamental fields of physics. So varied have these been that it is only possible to make a selection in a treatise of reasonable dimensions and such a selection has been made here.

The Principle of Coherence

When light is propagated as a wave form it is assumed that the medium through which the wave travels exhibits local periodic simple harmonic displacements. These repeat themselves in a time T, the period.

The source itself will for simplicity be considered to emit simple harmonic waves, the displacement y^1 of the oscillator being given at time t by an expression of the form:

$$y^1 = a \sin (2\pi/T)t$$

This is of course a purely classical concept, and indeed throughout this whole monograph quantum concepts will be avoided.

Now if we consider a point S distant x from the source, then the wave requires the time x/c to reach it, where c is the velocity of propagation. A vibrating element at this point S has for its equation of motion:

$$y = a \sin (2\pi/T)(t-x/c)$$

since the wavelength λ equals cT, this can also be written:

$$y = a \sin 2\pi(t/T-x/\lambda)$$

or alternatively one can write:

$$y = a \sin (2\pi/\lambda)(ct-x)$$

Clearly the phase at the point S depends upon the phase of the source. If light beams from two *independent* sources reach the point S there will be no fixed relation between the phases of the two light beams and it will not be possible to combine them to form stationary waves. Such light waves are termed *incoherent* and, of course, their intensities combine locally. On the other hand two light beams which arrive at S, say by different paths but are both radiated from *one* point of a source, are called *coherent* and such waves can superpose and produce visible interference effects because their *amplitudes* can combine, not their intensities. Actually, incoherent light waves may produce interference, but so rapid is the vibration frequency of visible light (about 10^{15} times a second) and so frequent are the phase changes from any source, that any interference effects arising will occur far too rapidly to be detected. With coherent radiations, on the other hand, *standing* waves can result, for any phase changes in the one beam originating in source variations are exactly duplicated in the other beam.

It is thus only possible to produce *stationary* interference effects (and only such effects are visible) by using two or more light beams which come from one point of a single source. When such coherent waves pass through a point, the vibrating medium at this point is subjected to the combined superposed effect of the two vibrations and under suitable conditions this leads to stationary waves; such waves are called *interference fringes*.

The Principle of Superposition

Let two coherent waves of different initial amplitudes a_1, a_2, but of the same wavelength λ, leave a point on the source and travelling by different path lengths, x_1, x_2, arrive at a point S where superposition takes place. *The amplitude Y of the resultant is the sum of the amplitudes y_1 and y_2 at the point* S. These are respectively:

$$y_1 = a_1 \sin (2\pi/\lambda)(ct - x_1)$$

$$y_2 = a_2 \sin (2\pi/\lambda)(ct - x_2)$$

Writing:

$$(2\pi/\lambda)(ct - x_1) = \phi$$

and

$$(2\pi/\lambda)(ct - x_2) = \phi + \theta$$

where $\theta = 2\pi(x_1 - x_2)/\lambda$, then

$$
\begin{aligned}
Y &= a_1 \sin \phi + a_2 \sin (\phi + \theta) \\
&= a_1 \sin \phi + a_2 \sin \phi \cos \theta + a_2 \sin \theta \cos \phi \\
Y &= (a_1 + a_2 \cos \theta) \sin \phi + a_2 \sin \theta \cos \phi \quad . \quad (1.1)
\end{aligned}
$$

The quantity θ is the *phase* difference at S and is merely 2π times the number of wavelengths in the path difference between the two beams reaching S.

To reduce this value of Y to a convenient form, use is made of the following identity. An expression such as $K = A \sin \phi + B \cos \phi$ can be written identically (by dividing and multiplying by $\sqrt{(A^2 + B^2)}$) as:

$$K = \sqrt{(A^2 + B^2)} \left\{ \frac{A}{\sqrt{(A^2 + B^2)}} \sin \phi + \frac{B}{\sqrt{(A^2 + B^2)}} \cos \phi \right\}$$

It will be seen from Fig. 1.1 that, in the right-angled triangle illustrated,

$$A/\sqrt{(A^2+B^2)} = \cos \epsilon \quad \text{and} \quad B/\sqrt{(A^2+B^2)} = \sin \epsilon$$

in which the angle ϵ is given by $\tan \epsilon = B/A$. Thus

$$K = \sqrt{(A^2+B^2)}(\cos \epsilon \sin \phi + \sin \epsilon \cos \phi)$$
$$K = \sqrt{(A^2+B^2)} \sin (\phi+\epsilon)$$

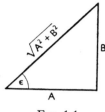

FIG. 1.1

By analogy from the equation (1.1) above for Y we can write:

$$A = (a_1+a_2 \cos \theta) \quad \text{and} \quad B = a_2 \sin \theta$$

giving $\quad Y = \sqrt{\{(a_1+a_2 \cos \theta)^2+(a_2 \sin \theta)^2\}} \sin (\phi+\epsilon)$

where $$\tan \epsilon = \frac{a_2 \sin \theta}{a_1+a_2 \cos \theta}$$

Simplifying the root term gives

$$Y = \sqrt{\{(a_1^2+a_2^2+2a_1a_2 \cos \theta)\}} \sin (\phi+\epsilon) \quad (1.2)$$

The effect at S is then such that the local vibrating element has a displacement with a period the same as that of the two component waves, but with a phase determined by the quantity ϵ and an amplitude determined by the term under the root.

Since the intensity of a simple harmonic vibration is proportional to the square of the amplitude, the *intensity* I at S is proportional to $(a_1^2+a_2^2+2a_1a_2 \cos \theta)$.

Considering first the simple case in which both combining waves have the same intensity, i.e. $a_1 = a_2$, then I has the value $2a_1^2(1+\cos \theta) = 4a_1^2 \cos^2 \theta/2$. Since $\cos \theta$ can vary from $+1$ to -1 it is clear that I varies from $4a_1^2$ to 0.

It will be recalled that $\theta = (2\pi/\lambda)(x_1-x_2)$ where (x_1-x_2) is the path difference between the waves arriving at S. Let this be denoted by d.

Since I is a maximum for cos $\theta = 1$, i.e. for $\theta = 0$, 2π, 4π, etc., then I is a *maximum* for $2\pi d/\lambda = 2\pi n$, i.e. for $d = n\lambda$ where n has all integral values.

Similarly I is a *minimum* for $\theta = \pi$, 3π, 5π, etc., i.e. for $d = (n+\frac{1}{2})\lambda$.

The simple physical meaning of this conclusion is that wave crests superpose and combine to a maximum when the path difference is a whole wave or an integral multiple thereof, whilst crests and troughs fall together to give zero effect when the path difference is an odd number of half-waves.

Hydrodynamical Analogy

To obtain coherence it has been postulated that the two combining waves come from the same point on the source and this must indeed be the condition in optics. It is, however, possible

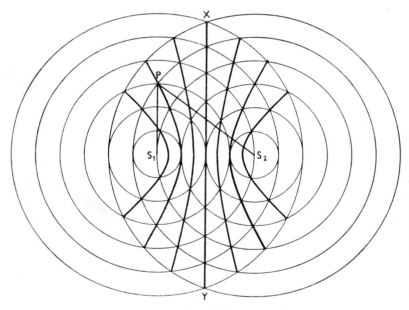

Fig. 1.2

to obtain coherence in a very useful hydrodynamical analogy in which there are two separate sources, yet they are still coherent. Such a case arises for example if two needles are attached to the prongs of a vibrating tuning fork and dipped into water, each creating circular waves which spread out over the surface and produce interference. For the two disturbing needles are strictly in phase thus standing waves are formed. This hydrodynamical analogy frequently holds closely in optics, for light interference often takes place between two waves reflected from slightly separated, or slightly inclined surfaces, the net effect being as if there were interference between a source and its coherent *virtual* image which is displaced by virtue of the reflection.

It will be found that this hydrodynamical analogy will prove very useful in considering the nature of the fringes formed under particular conditions. Fig. 1.2 shows how these arise. From S_1 and S_2 circular waves spread out on the surface of the liquid, strictly in phase. Let the circles represent crests, then where they intersect there will be a maximum. These maxima clearly lie on the heavy black curved lines.

Consider now these heavy lines only. At a point P there is a maximum when $S_2P - S_1P = n\lambda$. As the two sources are in phase the amplitude at P remains constant over time since its value only depends on distances, i.e. stationary waves appear.

The locus of points for a given value of n is represented by the equation: $S_2P - S_1P = $ constant. This is the equation for a hyperbola, of which S_1 and S_2 are the foci. The thick curves are hyperbolae and represent the localization of fringes, each fringe being defined by a specific value of n. Along the line XY the path difference is zero, i.e. n is zero, and this is termed *zero order*; successive "orders" arise for $n = 1, 2, 3$, etc.

Now imagine, instead of *plane* waves in two dimensions, spherical waves radiating in three dimensions from S_1S_2. Then to obtain the standing waves it is only necessary to rotate the pattern about the line S_1S_2, producing fringes which are hyperboloids of revolution obeying similar geometrical relations.

The Fringes on a Screen

The standing waves (fringes) given by two coherent effective light sources extend out into space and are called *non-localized* to distinguish them from other types of fringes which appear at one local position only. They can be seen either with a microscope or a telescope, or on a screen.

Suppose a small screen is placed in the position AB in Fig. 1.3, i.e. parallel to S_1S_2 (the line joining the effective coherent sources), then the intersection of the hyperboloids with AB gives the fringe pattern. If the screen-source distance is much greater than S_1S_2, the intersections will produce fringes on AB which are closely enough parallel equidistant lines.

If the screen is placed at PQ, with the two effective sources in line, then it is clear that the intersections of the hyperboloids of revolution with the plane will produce *circular rings*.

Suppose the screen at AB is moved round towards PQ and then round to the top of the diagram until it is once more parallel to

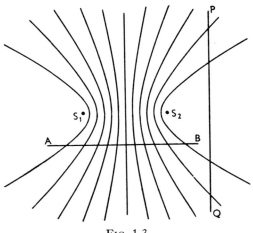

FIG. 1.3

S_1S_2. The fringes on the screen will pass successively through straight lines, arcs, circles, arcs, straight lines.

Although we shall discuss the shapes and positions of these fringes in much more detail later, it is emphasized that this simple description of the way in which the shapes of the fringes depend upon the positions of the screen relative to the line joining the sources is an important key-point in the explanation of a considerable number of interference experiments and will be frequently brought back into the discussion.

The Classification of Interference Effects

Optical interference phenomena will here be grouped according to four distinct attributes which can conveniently be called: (1) chromatism, (2) wave division, (3) fringe location, (4) multiplicity. Every interference phenomenon has an attribute belonging to each of these four classes and it will be found that the various complex phenomena to be described later can be fitted into this simple classification, which will now be considered briefly.

(1) *Chromatism*. Some interference effects are readily produced by white light, whilst others require an approximately monochromatic source of radiation. In the former case λ varies over a wide range, whilst in the latter λ is taken to be constant. "Monochromatic" radiation is merely a convenient figment since even single spectrum lines extend over an appreciable wavelength band. This "line width", under certain circumstances, imposes severe limitations on the interference experiments and will be discussed in detail later.

(2) *Wave division*. To satisfy the condition of coherence the only way by which coherent light beams can be obtained is to divide a single wave according to variants of two methods. The first method is called division of the *wave-front*, details of which will be given later. The second method involves *division of amplitude*, usually by partial reflection, as by a glass surface.

(3) *Fringe location*. It will be found convenient and helpful in classification to divide fringes according to their effective location. Fringes may appear to be (a) non-localized in the manner already described, (b) localized on a definite surface, often conveniently arranged to be effectively a plane, (c) at infinity, and seen only by a relaxed eye, or with a telescope set on infinity.

(4) *Multiplicity*. Many interference effects included in the above three classes can often be produced either with *two* light beams in which case the mechanism will be called *two-beam interference*; or with a number of related light beams, when it will be called *multiple-beam interference*. The multiple-beams require to be related specifically both in intensity and in phase, otherwise a confusion results. The use of many beams affects the widths and intensities of the fringes but usually not their location, shape, size or separation. It will, however, become apparent later that considerable advances result from merely altering fringe widths, and a good deal of discussion will be devoted to the treatment of the properties and applications of multiple-beam fringes.

CHAPTER 2

DIVISION OF WAVE-FRONT: DIFFRACTION

Historical

Before proceeding to consider in detail the various known facts about interference it is necessary to make an elementary survey of the phenomena of diffraction which also involve interference effects. Diffraction, which means the deviation from rectilinear propagation observed in waves, arises from the mutual interference interactions between the different parts of *one and the same wave-front*. The slight optical diffraction effects which are formed by a small obstacle or by a staight edge were first observed by Grimaldi in 1665, who noticed that shadows were not quite the same size as the anticipated geometrical shadows and were, furthermore, bordered by alternate narrow bright and dark fringes, very close to the edge and fading away rapidly on moving out from the edge. These observations were the cause of considerable speculation at the time and received a first qualitative explanation by Huygens who announced his celebrated secondary wavelet construction in 1678 largely as a result of Grimaldi's discoveries.

Huygens' Secondary Wavelet Theory

Huygens' geometrical construction was designed to explain rectilinear propagation in accordance with the following. A light source is considered to produce a spherical wave. Each point on the wave-front is considered to become the source of secondary wavelets. These interfere destructively everywhere except at the resultant tangential envelope. This will be made clear by Fig. 2.1. Let light diverging from the source Z pass through an aperture AB. Let the spherical wave-front after an interval be CD, and let each point on CD become a centre of disturbances. Construct a series of circles of equal radius ct, then after time t the wave has reached the envelope PQ.

Clearly the lines ZAX and ZBY define the propagation which is thus rectilinear in the sense of geometrical ray tracing.

The wavelets originating on CD destructively interfere in the region PQRS except on the envelope PQ which propagates forward (although Huygens side-stepped the difficulty of explaining away the absence of the dotted *rear* envelope RS which would be projected backwards). However, the theory clearly indicates that some light should appear *outside* of the geometrical region ZPQ because the circles drawn on the regions near RP and SQ with C and D as centres are not fully compensated. The smaller AB the bigger is this "side-spread" effect and the greater the amount of light outside the geometrical limit. Thus in a qualitative way diffraction is accounted for.

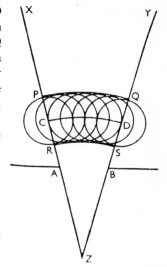

Fig. 2.1

Young's Slits Experiment

Although the analysis of diffraction effects was first formally given by Fresnel in 1815, the celebrated interference experiment made by Thomas Young in 1802 is essentially based on diffraction. This experiment was to prove that Newton's corpuscular theory

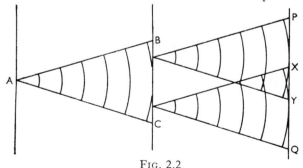

Fig. 2.2

of light was inadmissible, and was made with crudely mono-chromatic radiation in the following manner.

Light diverges from a narrow slit A as illustrated in Fig. 2.2. B and C are two fine slits, close together and parallel to A. The two elements of the circular wave which pass through B and C are coherent.   Since B and C are narrow the light diffracts out from each effectively over a cone.   Where these two cones overlap, interference takes place and if a screen PQ is held in the path interference fringes are seen in the overlap region XY.

Young proved that this was an interference effect by the simple expedient of covering up the slit B, when the fringes vanished. They were thus not inherently due to diffraction but to interference.   However, since the conical spread of the beams is due

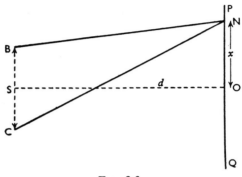

FIG. 2.3

to diffraction the interference effect is not quite simple, the *fringe intensities* being modified in a periodic manner owing to diffraction. If this refinement be disregarded one can derive the approximate form of the fringe distribution.

Let the two slits (Fig. 2.3) B and C be a distance s apart and let the screen PQ be distant d from BC.   At the point O the beams from B and C arrive in phase since there is zero path difference. This point must be bright.   Let the nth bright fringe appear at N where $ON = x$.   Then by definition $CN - BN = n\lambda$.

Now $CN^2 - BN^2 = \{d^2 + (x + s/2)^2\} - \{d^2 + (x - s/2)^2\}$

$$= 2xs$$

Thus $CN - BN = 2xs/(CN + BN)$

If s is small compared with d then $CN + BN$ is approximately $2d$ (this is the case in practice).

Then $CN - BN = xs/d$

But $CN - BN = n\lambda$

Thus $$xs/d = n\lambda$$

i.e. $$x = n\lambda d/s$$

The $(n+1)$th fringe will appear at x_1 where $x_1 = (n+1)\lambda d/s$.

Hence the separation Δ between a pair of fringes is

$$\Delta = x_1 - x = \lambda d/s \quad . \quad . \quad . \quad . \quad (2.1)$$

To this approximation the fringes are equally spaced and with a separation Δ proportional to λ and to the screen distance d, and inversely proportional to the slit separation. More precisely the fringes are not quite straight lines, but hyperbolae. However, their separations change only slowly on moving from zero order.

The fringe separation is small, even when the slits are close together; for suppose the slit separation is 1 mm, then on a screen distant 1 metre, for a wavelength $\lambda = 5 \times 10^{-5}$ cm (green light), the fringe separation is 0.5 mm. Young's experiment was therefore not easy to carry out with the meagre equipment then available. However, his observations sufficed not only to vindicate the wave theory but to give a value for λ.

An important quantity, which will often be used later, is the *angle subtended at the source by a pair of fringes*. This is Δ/d and, as seen from equation (2.1) is equal to λ/s.

Chromatism of the Fringes

The separation between fringes is proportional to λ, hence the scale of the fringe pattern depends on the wavelength. With a monochromatic source the fringes are equidistant and Fig. 2.4 (a), (b), (c), show the relative scales of red, green, and blue fringes ($\lambda\lambda$ 6000, 5000, 4000), O being the zero order. Below at (d) is shown the appearance if these three systems are superposed. It will be seen that there is a bright central image, flanked by two dark fringes B, B′, but beyond that the fringes run into each other and merge to form white light. It is clear that with white light very few fringes other than the zero order will be visible and those that are seen will be coloured. This is an important characteristic, for it permits identification of the zero order. With monochromatic radiation (either (a) or (b) or (c)) all the fringes are alike and it is not possible to distinguish one order from the next. With the white-light fringes, on the other hand, zero-order identification is unambiguously defined.

If a thin piece of mica of thickness t and refractive index μ is introduced into the path BN of Fig. 2.3, the optical path length increases by $(\mu - 1)t$. Thus O is then no longer the position for which there is equal path difference. The zero-order white-light fringe then moves up towards P, say a distance y. If the number of fringes, n, of any selected wavelength λ, in the distance, y, is measured then $(\mu - 1)t = n\lambda$. It is thus possible to evaluate

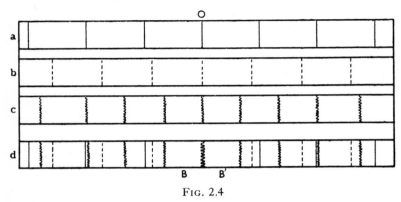

FIG. 2.4

either μ or t if one is known. *This measurement cannot be carried out on monochromatic fringes alone* for then the number of fringes displaced cannot be determined. Such a failure to identify the order displacement of fringes is quite a frequent condition in many interference systems if a single monochromatic radiation only is used.

Diffraction by a Rectangular Aperture (Slit)

Diffraction with diverging light beams is called Fresnel diffraction. Such effects can be complex, but for our immediate purpose another type of diffraction is of more importance and this is the type called Fraunhofer diffraction. This takes place when a plane wave (i.e. parallel light beam) is incident on to an aperture, the resulting diffracted light then being brought to a focus by a lens. The Fraunhofer diffraction effect appears at the lens focus. An elementary analysis is included here, for the conclusions drawn are needed frequently in interferometry. The optical arrangement is shown in Fig. 2.5. A parallel light beam falls at normal incidence on to a slit and the light emerging is diffracted and appears to issue with an angular spread. The intensity in any arbitrary direction θ is evaluated below.

Let the width of the aperture be e, then the two extreme rays coming to the focus F in any diffracted direction θ have a path difference $e \sin \theta$. The number of waves is thus $(e \sin \theta)/\lambda$ and the *phase* difference between these beams is $2\pi.(e \sin \theta)/\lambda$.

Each element of the wave-front AB has an effect at F and there results superposition and interference due to their separate contributions. The effect at F due to these elements will now be calculated.

We can imagine the wave-front AB divided off into a very large number of equal elements, all of the same frequency and the same amplitude, and with the same path difference between each adjacent element. On the wave-front AB the phases of each element, on going from A to B, increase in arithmetic progression.

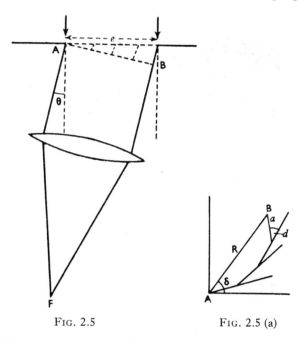

Fig. 2.5   Fig. 2.5 (a)

Now, as is well known, simple harmonic vibrations of *the same period* can be compounded by the parallelogram (or polygon) of vectors, and if we represent each element as a vector element strip, of which, for example, four are shown between A and B, the equal angle increments representing the equal phase increments, then the amplitude and phase of the resultant is given by the length of AB and the angle it makes with the horizontal axis (Fig. 2.5 (a)).

Let a be the amplitude of each element, d the common phase difference, R the amplitude of the resultant AB, and δ its final phase value, then resolving in the horizontal and vertical directions gives:

$$R \cos \delta = a(1 + \cos d + \cos 2d + \ \dots \ ) \qquad . \quad (2.2)$$

$$R \sin \delta = a(\sin d + \sin 2d + \ \dots \ ) \qquad . \quad . \quad (2.3)$$

Multiply both sides of both equations by $2 \sin d/2$ giving, from (2.2),

$2R \cos \delta . \sin d/2$
$= a\{2 \sin d/2 + 2 \cos d \sin d/2 + 2 \cos 2d \sin d/2 + \ \dots \ \}$
$= a\{2 \sin d/2 + (\sin 3d/2 - \sin d/2) + (\sin 5d/2 - \sin 3d/2) + \ \dots \ \}$

Since the same terms but with opposite signs occur in the brackets this reduces to the first and last terms only, i.e. to:

$$2R \cos \delta . \sin d/2 = a\{\sin d/2 + \sin (n - \tfrac{1}{2})d\}$$

i.e. $\quad 2R \cos \delta . \sin d/2 = 2a \sin nd/2 . \cos (n-1)d/2 \quad . \quad . \quad (2.4)$

in which n is the total number of elements involved.

In a similar way, from equation (2.3) one obtains:

$$2R \sin \delta . \sin d/2 = 2a \sin nd/2 . \sin (n-1)d/2 \qquad . \quad . \quad . \quad (2.5)$$

Square and add (2.4) and (2.5), from which

$$R^2 \sin^2 d/2 = a^2 \sin^2 nd/2$$

$$R = \frac{a \sin nd/2}{\sin d/2} \qquad . \quad . \quad . \quad . \quad (2.6)$$

Also by dividing (2.5) by (2.4) one gets:

$$\tan \delta = \tan (n-1)d/2$$

i.e. $$\delta = (n-1)d/2 \qquad . \quad . \quad . \quad . \quad (2.7)$$

Let n become very large, then correspondingly a must become very small, but the finite product na represents the total vector length AB in Fig. 2.5 (a). Let na be equal to A. In the same way as n increases d diminishes, but the finite product nd represents the phase difference between the beams at the extreme ends A and B. Let this extreme phase difference nd be called 2α. Since n is large then $\delta \ (= (n-1)d/2)$ is effectively equal to α. Hence (2.6) can now be rewritten as:

$$R = \frac{a \sin \alpha}{\sin \alpha/n}$$

But since α/n is now small, then closely enough:

$$R = \frac{a \sin \alpha}{\alpha/n}$$

$$= \frac{na \sin \alpha}{\alpha}$$

$$= \frac{A \sin \alpha}{\alpha}$$

R is the amplitude hence the *intensity* is given by $(A^2 \sin^2 \alpha)/\alpha^2$.

To find how R varies with α, differentiate and equate to zero to obtain maximum and minimum values. This gives:

$$\frac{d}{d\alpha} \cdot \frac{\sin \alpha}{\alpha} = \frac{\cos \alpha}{\alpha} - \frac{\sin \alpha}{\alpha^2} = 0$$

or $$\alpha - \tan \alpha = 0$$

$$\therefore \ \alpha = \tan \alpha$$

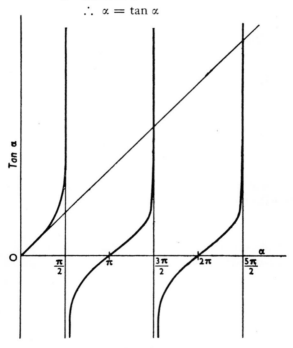

FIG. 2.6

The solution to this equation can be got either from tables in which those angles equal to their tangents are found, or alternatively graphically as follows (Fig. 2.6): Draw a graph of the variation of tan α with α and through the origin draw a straight line making an angle of 45° with the axes. The intersections of this line with the tangent curves are the roots of the equation. As shown schematically in Fig. 2.6 the roots are at values of α which are zero and then approach closer and closer to the values $3\pi/2$, $5\pi/2$, $7\pi/2$, etc. The square of these values gives the intensity maxima.

The *minima of intensity* (zero) all occur precisely at values of $\alpha = n\pi$, since the formula then gives $R = 0$ as $\sin \alpha = 0$. At the value $n = 0$, however, the quantity $(\sin \alpha)/\alpha$ becomes 1 and $R^2 = A^2$.

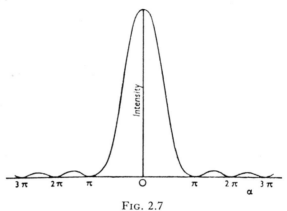

FIG. 2.7

The maxima are not at the exact odd $\pi/2$ values but are quite close, the following being more correct values:

$$\alpha = 0, \quad 1\cdot43\pi, \quad 2\cdot46\pi, \quad 3\cdot47\pi, \quad 4\cdot48\pi$$

To obtain the intensities of the maxima it is sufficiently close to use the half-integral values:

$$\alpha = 0, \quad 3\pi/2, \quad 5\pi/2, \quad 7\pi/2, \quad \text{etc.}$$

Substituting in $R^2 = (A^2 \sin^2 \alpha)/\alpha^2$ gives the ratios of successive maxima (i.e. taking A as unity) as:

$$1 : (2/3\pi)^2 : (2/5\pi)^2 : (2/7\pi)^2 \ldots$$

i.e. $1 : 1/21 : 1/61 : 1/120 \ldots$

A plot of the way in which R^2 varies with α is shown schematically in Fig. 2.7. Because the secondary maxima are not quite equidistant there is a slight asymmetry.

Physical Interpretation

We shall now consider the physical interpretation of this diffraction curve. The quantity plotted is the intensity against α. But 2α, the phase difference between the extreme ends of the beam is by definition $2\pi.(e \sin \theta)/\lambda$, hence $\alpha = \pi.(e \sin \theta)/\lambda$.

The horizontal scale of the curve is therefore determined principally by the value of e, the width of the aperture, since λ does not vary a great deal for the complete range of visible light.

The position of the first minimum occurs at $\pi = \alpha$, i.e. at $\pi = \pi.(e \sin \theta)/\lambda$, thus at $e \sin \theta = \lambda$. In this, θ is the angle of diffraction, namely the angle of deviation from the normal. Since λ is small, say 5×10^{-5} cm, then, if e is large, θ will also be small, hence the angle of diffraction for the first minimum is so small that effectively a geometrical image forms when e is large, with extremely narrow faint secondary maxima very close to the edges.

However suppose e approaches λ in value, then $\sin \theta$ becomes near unity, which means physically that the light effectively spreads out almost hemispherically from the aperture. This is in harmony with Huygens' concept of secondary wavelets, for the aperture has effectively now become a source of spherical waves. The treatment breaks down for $e < \lambda$, which would make $\sin \theta > 1$.

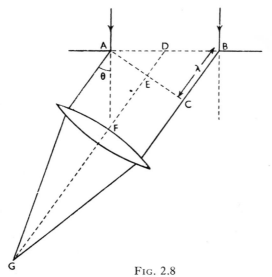

FIG. 2.8

It is clear that for narrow slits the phenomena are intermediate between those two just considered, and the narrower the slit the larger the angular separation between the minima and maxima.

That the position of the first minimum occurs at $e \sin \theta = \lambda$ is alternatively easily derived by the following method of Rayleigh (Fig. 2.8). Consider diffraction in a direction θ selected such that the retardation BC between the first and last ray is λ. For the middle (dotted) ray DEF the retardation DE is $\lambda/2$. For every point between AE, there is a corresponding point between EC just $\lambda/2$ behind, hence in this direction θ the whole effect of AE is completely annulled by the effect of EC. The intensity at G is thus zero, and therefore this is the direction of the first minimum. As BC $= e \sin \theta$, then $e \sin \theta = \lambda$ gives the value of θ for the first minimum, in agreement with the more formal deduction given previously.

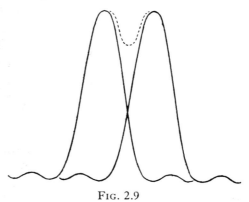

FIG. 2.9

Resolution of Two Slit Images

Of some considerable importance to interferometry is the manner in which the above considerations lead to the conditions for the resolution of the images of two slits which are close together. As a lens has a finite aperture it does not produce completely sharply defined geometrical images of the two slits but images which have a definite diffraction width, each being flanked by secondary maxima. The finite diffraction width sets a limit to the ability to separate the images of two slits close to each other. It has been found convenient to use the Rayleigh criterion (given below) for resolution.

Rayleigh arbitrarily assumes that two diffraction images, from two similar slit sources, can be resolved when the main maximum of the one falls on the first minimum of the second. This case of resolution is shown in Fig. 2.9.

From the symmetry, the point of intersection of the two curves occurs at $\alpha = \pi/2$, since the first minimum is at $\alpha = \pi$. Substituting this in $R^2 = (\sin^2 \alpha)/\alpha^2$ gives $4/\pi^2 = 0\cdot405$ for the intensity at the point of intersection. The dotted curve shows the combined result of adding the two diffraction curves. The dip in the saddle has the intensity $2(0\cdot405)$, i.e. $0\cdot81$. Rayleigh assumes that the eye can just detect a saddle-back with this minimum of some four-fifths the intensity of the maximum. It is thus mathematically very convenient to adopt for resolution the superposition of one maximum on the first minimum of the other. This condition often arises and will later be met with frequently. Visual resolution is in fact usually slightly better than the Rayleigh criterion.

Angular Resolution of a Mirror

It will be shown later that by special interferometric methods it is possible to measure smaller angles than can be normally measured either by rotating mirrors or goniometers. The smallest measurable angle through which the rotation of a suspended mirror (e.g. a galvanometer mirror) can be observed is fixed by diffraction considerations. When a parallel beam of light from say a slit source is reflected by a mirror, then the latter is equivalent to an aperture having the dimensions of the mirror since one can imagine a virtual source behind the mirror sending out a plane wave which appears to pass through an aperture of the mirror dimensions. The image obtained at the focus of a lens will therefore have the Fraunhofer diffraction pattern of an aperture. The spread of this diffraction pattern determines the smallest deflection measurable, and one can apply the Rayleigh criterion for resolution to obtain this

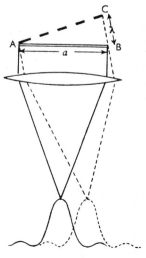

Fig. 2.10

angle, as is shown in Fig. 2.10 in which AB is the mirror, which then rotates to AC. The limiting resolution condition requires the maximum of the one image from AC to fall on the first minimum of the image from AB. To find where this is it is only necessary to make BC $= \lambda$, as has already been shown. If the aperture of the mirror is a then the angle CAB $= \lambda/a$. Hence λ/a is the smallest angle a mirror can resolve. It is also the smallest angle (say on a crystal surface) that can be resolved with a goniometer, and in this case the larger the crystal faces, the smaller the angle that can be resolved.

Considering as a typical case a galvanometer mirror which is 5 mm across, then for $\lambda = 5 \times 10^{-5}$ cm the limit angle is $(5 \times 10^{-5})/(5 \times 10^{-1})$ radian, i.e. 10^{-4} radian, so that with a screen distant 1 metre a deflection of 0·1 mm can just be resolved. Since this is easily seen with a simple magnifying lens there is no gain in projecting the beam over much longer distances in the hope of increasing sensitivity, for this merely produces a broader image.

Angular Resolution of a Telescope

It is necessary for the appreciation of some interference experiments to be familiar with the diffraction effects which restrict the angular resolution of a telescope. The conditions are practically the same as those arising in the case of the mirror. See Fig. 2.11.

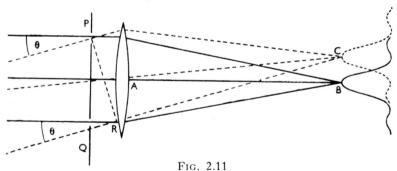

FIG. 2.11

Suppose a lens of focal length f $(= \text{AB})$ has before it an aperture of a cm $(= \text{PQ})$. A normally incident parallel beam from say a star produces a Fraunhofer diffraction image at B. Now consider light coming in a direction θ from another star, forming an image at C. This will just be resolved if QR $= \lambda$ for then C will fall on the first minimum of B.

From triangles PQR and ACB, which are similar, $\lambda/a = BC/f = \theta$. Thus, as in the mirror, the angle that can be resolved is λ/a. The size of the resolvable image BC is $f.\lambda/a$.

This calculation applies only to a rectangular aperture. In general, telescopes have a circular aperture, in which case, as first shown by Airy by integrating across the circular diaphragm, the limit angle of resolution both for a circular mirror and a telescope is modified to $1\cdot22\lambda/a$. If instead of viewing a slit image (which

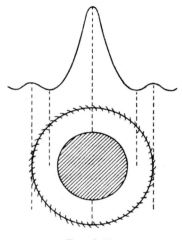

FIG. 2.12

is bordered by the secondary diffraction bands parallel to the slit) observation is made on a point source, then the diffraction pattern is a disc, called the Airy disc, of diameter $1\cdot22\lambda/a$ surrounded by rings (haloes) at the distances corresponding with the successive secondary images for the slit, as shown in Fig. 2.12. If a section is taken across the disc, then the intensity distribution is like that of a rectangular aperture so that similar conditions as to resolution apply for disc images as for slit images.

CHAPTER 3

LIGHT SOURCES

Introduction

For the chromatic fringes, which are formed by white light, any convenient white-light source may be used. According to the light intensity required one can employ an incandescent lamp, pointolite or high-current carbon arc. The pointolite is a particularly bright stable source and is ideal for many purposes.

For "monochromatic" fringes the nature of the particular investigation determines the degree of "monochromatism" required. This can vary from the broad wavelength band transmitted by a coloured glass filter to the sharp bright monochromatic line sources which have been developed for special spectroscopic purposes. Advances in precision measurements using interference have been closely related to the development of special monochromatic sources, and there has been a mutual interaction between the evolution of sources and interferometric appliances, an increase in the quality of the one leading to a search for an improvement in the other. This process is still actively going on.

Monochromatic Line Sources

Monochromatic sources are of two kinds, according to their purpose. In the first type the source is merely a convenient light emitter used to produce some interference effect. The degree of monochromatism depends upon the conditions of observation, and a frequent condition is that the source must needs be bright. The nature of the light itself is in certain experiments of no interest, although its quality affects the observations in a number of ways to be discussed in detail. When the path *difference* is considerable, i.e. when the order of interference is high, say >200 000 waves then the monochromatism must be very high. When, however, the path difference is only a few light waves, a broadened line can be tolerated, indeed a close group of separate lines can be considered to be effectively monochromatic. For example the separation of the two sodium D lines (6 Å) is often of no significance.

In an entirely different class of interferometric studies, interference methods are applied to the examination of the properties of the light radiation itself, and in this case it is often necessary to secure as high a degree of monochromatization as is possible. In the following sections some of the devices developed for these various interference purposes will be briefly discussed. The factors affecting the widths of spectrum lines will first be reviewed.

Line Widths

A number of causes contribute to the widths of spectrum lines and these will be briefly considered in order to discuss how broadening effects can be avoided, when necessary. The main broadening factors are respectively: (a) Doppler width, (b) pressure broadening, (c) resonance width, (d) Stark broadening, (e) self-reversal, (f) hyperfine structure.

(a) *Doppler width.* Doppler width results from the thermal motions of atoms or molecules. Suppose all the atoms in a gas emit a strictly monochromatic radiation of wavelength λ, then

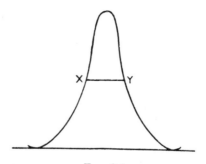

FIG. 3.1

those moving *towards* an observer with velocity v *appear* to emit a wavelength $\lambda - d\lambda$, where $d\lambda$ equals $\lambda . v/c$, c being the velocity of light. Also those moving *away* from an observer with the same velocity give out radiation which *appears* to the observer to have the wavelength $\lambda + d\lambda$. Because the atoms in a gas move with velocities which fall on a Maxwell distribution curve the result is the production of a broadened spectrum line the shape of which, when intensity is plotted against wavelength, has a Maxwell distribution as shown schematically in Fig. 3.1.

It is usual to define the broadening by the half-width $w = XY$, which is the width at half the intensity of the peak. It can be shown that this width in Ångström units is

$$w = 0 \cdot 7 \times 10^{-6} \times \lambda \sqrt{(T/M)}$$

in which T is the absolute temperature and M the atomic weight of the emitting atom.

To reduce Doppler width with a given source one must decrease the temperature. To take an example, the Doppler half-width of the yellow sodium lines in a hot-arc source at $3600°K$ is $0 \cdot 052$ Å. If the temperature be cooled to that of liquid air ($100°K$), the width is reduced (but only by a factor of 6) and is approximately $0 \cdot 009$ Å.

If one considers the green mercury line, however, the greater mass of the emitter has an effect and the half-widths at the above two temperatures become approximately $0 \cdot 016$ and $0 \cdot 003$ Å respectively.

For very many observations any of the above line widths is sufficiently monochromatic, but for some very special investigations even a width of only $0 \cdot 003$ Å is detrimental.

(b) *Pressure broadening.* Pressure broadening, which is due to collision bombardment, depends upon the temperature, density and nature of the gas wherein the emitting atoms are situated. Thus if the sodium lines are excited in nitrogen gas at a pressure of 1 atmosphere, the pressure broadening amounts to $0 \cdot 1$ Å and is approximately proportional to pressure. This is twenty times as great as the Doppler width at even $3600°$, hence to obtain narrow lines a vacuum source must be used as this virtually eliminates pressure broadening. High-pressure mercury-arc sources can be used to advantage in many experiments, but, since the pressures may even exceed 50 atmospheres, line widths are then great. In such a source pressure broadening alone might be expected to be of the order of 2 Å or more. Even this width is insignificant in some observations, as will be shown later.

(c) *Resonance broadening.* This is a particularly intense form of pressure broadening and arises when the pressure is *self*-vapour pressure. Thus the line-width pressure broadening produced on sodium lines by a pressure of 1 cm of *sodium vapour* may easily be 100 times that produced by a 1 cm pressure of nitrogen gas. Different spectrum lines show different degrees of sensitivity to this effect. Because of *self*-pressure resonance broadening the width of the green line in a high-pressure mercury-arc usually appreciably exceeds 2 Å and can even approach 50 Å.

(d) *Stark broadening*. This arises through Stark effect, the displacement by an electric field of atomic energy levels. Lines differ enormously with respect to their behaviour in electric fields. On the yellow sodium lines the broadening is less than 0·001 Å for each thousand volts per cm of field, whereas on the blue hydrogen line it is 0·5 Å for the same field strength. With large field strengths considerable Stark broadening arises in some cases, and then suitable precautions require to be taken.

(e) *Self-reversal*. This effect is mainly noticed with resonance lines, such as the yellow sodium D lines. The radiation passing through its own vapour is partially absorbed. The centre of the line is preferentially absorbed relative to the wings, and the line profile flattens, with consequent effective increase in the defined width. The reversal width depends upon the length of the path and the gas pressure and is very sensitive to temperature. With vapour at a pressure of 0·01 mm (temperature 300°C) the D lines from a 20-cm deep source have a width of about 0·6 Å. There is still some broadening even at a vapour pressure of 0·001 mm.

At temperatures above 300°C complete reversal appears, that is the centre of each line is so strongly absorbed that the line appears to be a spurious doublet. This effect can often be noticed in interference experiments made with sodium-flame sources.

(f) *Hyperfine structure*. A broadening effect which cannot be avoided (apart from exceptional cases to be indicated) owes its origin to the hyperfine structures of spectral lines. It is known that many of the strongest lines in many spectra are not simple lines but when examined with high resolution are found to possess complex hyperfine structures. Indeed these structures are best studied by interferometric means, for under specified conditions the interference effects can be made to reveal line widths and line complexities.

These hyperfine structures are usually on quite a small scale (hence their name), and if, as is often the case, they are not separately resolved their combined effect is to give a spectrum line an effective broadening. There is some considerable variation in the spread of hyperfine structure patterns which may variously extend over as little as perhaps 0·001 Å or as much as perhaps 1 Å.

Hyperfine structures arise from two quite distinct causes, namely: (a) nuclear spin, (b) isotopy. If an atomic species has an odd atomic weight then it is known that this atomic nucleus has an angular moment (nuclear spin). Such a moment interacts with the magnetic fields of the atomic electrons and as a result most spectrum levels split up. A "line" transition between two levels is no longer single but becomes a line complex and this is the spin hyperfine structure. Such a structure is an inherent property of the spectrum line in question and cannot be influenced or eliminated. The individual hyperfine-structure components can be subject to all the broadening influences already discussed.

If an atomic species consists of a mixture of isotopes, then strictly speaking each isotope produces its own separate lines. However, these often fall very close together and are only distinguishable in certain instances. When they are so distinguished the line in question exhibits an "isotopic" hyperfine structure. It is possible to modify this by separating out single isotopes from isotopic mixtures.

Typical Sources

It is of some interest to review the history of monochromatic sources. Newton carried out his classic investigations on Newton's rings by isolating a narrow band from the sun's spectrum with the aid of a slit. Fresnel's investigations on interference were made with white light filtered by a piece of red glass, a very crude form of monochromatization, since the band was possibly 1000 Å wide. The first to employ an effectively monochromatic source, in the modern sense, was Brewster, who, in 1828, made use of sodium chloride in an alcohol flame, employing thus the sodium D lines 5890, 5896 Å. The sodium flame, replaced later by the Bunsen burner, impregnated with sodium chloride, remained for many years the most important of all monochromatic sources. There was little improvement upon the flame for some sixty years, after which the use of the Geissler-tube discharge through hydrogen began to be adopted.

The first really important step in producing bright steady monochromatic sources was taken by Michelson in 1889 who excited metal vapours at a pressure of a small fraction of a

millimetre by means of an induction coil.   The most important of
these sources was the cadmium lamp illustrated in Fig. 3.2.   This
contained cadmium at a temperature of 300°C and a low pressure.
When excited with some 2000 volts and a current of a few milli-
amperes, using an induction coil, four bright lines are emitted,
of which the red line at 6438 Å is one of the sharpest lines yet

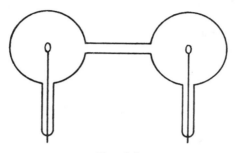

FIG. 3.2

observed and its wavelength was long the standard of length for
the whole of physical measurement.

The modern version of the cadmium lamp consists of a discharge tube such as the Osira type containing cadmium.   Such
tubes contain argon and cadmium metal.   A discharge of perhaps
half an ampere is passed and, on warming up, a brilliant cadmium
emission results.   Such tubes have a cross-
section of at least 1 sq. cm, and it is important
that the current density should not be excessive
otherwise line broadening results.   Thus con-
strictions and capillaries are to be avoided.

In 1892 Arons invented the mercury vacuum-
arc and for many purposes this is an excellent
interferometric source.   A simple type is shown
in Fig. 3.3.   It is an inverted U-tube, limbs
being filled with mercury up to the junction.
It is supplied with say 230 volts d.c., and on
tilting slightly an arc breaks across the mercury
columns.   The upper bulb acts as an air-cooled
condenser, and evaporated mercury collects
there and runs down to the limbs.   With a
2-cm bore tube such an arc normally requires
about $3\frac{1}{2}$ amperes to maintain itself.   Very

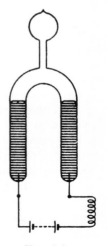

FIG. 3.3

much sharpened lines are produced if the current is cut to 1 ampere and this can be achieved by having a massive inductance in series with the lamp.   This stabilizes the arc, maintaining it even when considerably under-run.

The emission from such a source is brilliant and consists, in the visible, *largely* of a characteristic group of lines which are: a yellow pair 20 Å apart at 5770, 5790 Å, an intense isolated green line at 5461 Å, and a strong triple violet group near 4358 Å. These three line-groups are so far apart that cheap and efficient filters are available for separating them and this is a considerable advantage.   The green line in particular is most intense.   It has, however, a number of very close satellite components (hyperfine structure) which reveal themselves under certain conditions.   The individual hyperfine-structure components are often quite sharp.

Within recent years a notable development has been the appearance of commercial high-pressure mercury sources.   These are of two kinds.   In one the source is a narrow tube perhaps 5 cm long and 1 cm bore, made of quartz and with tungsten electrodes. A small amount of argon and a pellet of mercury are enclosed.   A discharge is passed and the temperature rises until the pressure of the mercury is several atmospheres.   Intensities of the order of 100 000 stilb are attained.

The lines are very broad, but often this is of little consequence. The particular importance of the source is the great intensity, which permits special methods to be developed in precision interferometry.

The most superior source from the point of view only of narrowness and approximate monochromatism is the atomic beam. In this a jet of atoms emerging from a pair of parallel slits is excited into emission.   Since all the atoms are moving practically in one direction, if the beam is viewed in a direction at right angles to the line of motion there is no effective resolved component of velocity in this direction of sight.   Thus there is no effective Doppler broadening and line widths of the order of the remarkably low value of 0·001 Å can be obtained for the red cadmium line. The lines so emitted are weak but it is of some interest to point out that with such radiations one should still obtain fringes with the order of interference equal to $6\frac{1}{2}$ million, which represents a path difference of more than 4 metres.

A recently developed source for precision interferometry has now been produced.   If gold (atomic weight 197) is neutron-irradiated in an atomic pile it converts to mercury of atomic weight 198. With this reaction some milligrams of this single isotope of mercury have been isolated and with it vacuum discharge tubes have been made. Excitation leads to emission of the mercury lines *with no hyperfine structure* whatsoever, since the single isotope 198 has no nuclear spin. Such sources have already found useful application, but have only a moderately short life.

CHAPTER 4

INTERFERENCE WITH DIVISION OF WAVE-FRONT

Fresnel's Mirrors

Although Thomas Young's interference experiment did in fact establish the correctness of the wave theory, yet many objections were raised on account of the fact that diffraction from slits was involved and as a criticism it was maintained that Newton's "inflexion" at an edge might still be invoked to explain the observations. There was not much basis in this view, but to silence objections Fresnel devised a series of experiments in which interference effects were produced without dividing the wave-front by diffracting slits. The simplest of these is the Fresnel *two-mirror* experiment (so called because of a later experiment involving *three mirrors*).

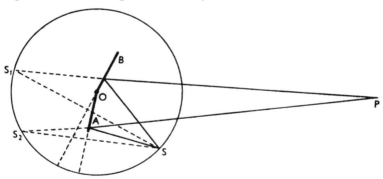

FIG. 4.1

In discussing Young's experiment it was shown that the smallness of the wavelength of light makes it necessary for the two slit sources to be close together if the fringes are to have a separation which can reasonably easily be seen with low magnifications. Fresnel, in 1816, produced this condition of closeness by reflecting light from a single-slit source S by two plane (front silvered) mirrors touching each other and very slightly inclined to one another. The arrangement is shown in Fig. 4.1. The mirrors

32

A and B produce images of S at S_1 and S_2 which are close together if the angle between A and B is small. These two virtual sources produce interference fringes on a screen at P. (To simplify the diagram a single ray is shown, instead of a diverging cone.)

From the geometry S, S_1 and S_2 lie on the circumference of a circle the centre of which, O, lies on the edge of contact of the mirrors.

If OS, the radius, $= a$, and OP, the distance to the screen $= b$, then if ω is the angle between the mirrors, the angle $S_1SS_2 = \omega$ and $S_1OS_2 = 2\omega$, since both stand on the same chord S_1S_2. Thus $S_1S_2 = 2a\omega$. It has been shown that two sources, s apart, produce fringes separated by a distance $\varDelta = \lambda d/s$ on a screen distant d. Since the source separation is $2a\omega$ and the screen distance $(a+b)$, the fringe separation is $\lambda(a+b)/2a\omega$.

The fringes extend into space and are thus non-localized. They can be seen on a screen or with either a microscope or telescope and can be produced with white light or with monochromatic light.

Fresnel's Biprism

Fresnel then designed a further experiment using a biprism.

In Fresnel's biprism (1826) a slit source is placed at a short distance a behind a symmetrical prism of which the two base angles are very small, perhaps $\frac{1}{3}°$, the slit being parallel to the prism edge.

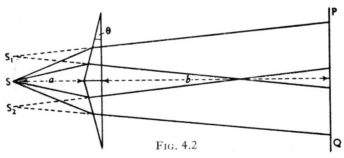

FIG. 4.2

Light from S (Fig. 4.2) is refracted by the two prisms to form two virtual images S_1 and S_2, which produce interference on the screen PQ distant b from the prism.

The angular deviation produced by a prism of refractive index μ and of small angle θ is $(\mu-1)\theta$ so that the linear distance SS_1 is, closely enough, $a(\mu-1)\theta$. This leads to fringes on a screen PQ, which is distant $(a+b)$ from S_1 and S_2, the fringe separation being $(a+b)\lambda/2a(\mu-1)\theta$. To make the image separation S_1S_2 small, S should be close to the prism. θ can be measured by setting the prism on a spectrometer or goniometer table.

In practice the distance S_1S_2 can be derived as shown in Fig. 4.3 without a knowledge of μ or θ.

A lens L is interposed to give an image of the virtual sources S_1S_2 in an eyepiece at P. If the distance S_1S_2 is x, then the image length at P is magnified and is $x_1 = x . Y/X$, where X and Y are the distances shown. The distance PL $= Y$ is measured. The lens is moved into its conjugate position L$'$ and the diminished image $x_2 = x . X/Y$ measured, as well as the distance PL$' = X$. Then $x_2x_1 = x^2$, and the distance between the source and the screen equals $X+Y$.

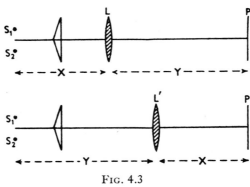

Fig. 4.3

Thus fringe separation is $\varDelta = (X+Y)\lambda/x$ and no knowledge of μ or θ is required.

Although the biprism is mainly of academic interest, yet it has found at least one application, for in 1870 Töppler and Boltzmann used it to examine the changes in air density at the end of a transparent vibrating organ pipe. Of the two interfering beams one passed through the pipe and the other outside of it, and changes of refractive index caused by the vibration could be observed when the illumination was stroboscopically controlled with a tuning fork.

Split-lens Interference Effects

In the experiments of Fresnel the interference is produced by two *virtual* images. In 1858 Billet devised a simple experiment in which the interference is caused by two *real* images produced by a lens. This will now be reviewed, since it is intended to discuss later in some detail the formation of fringes from real sources produced by a lens.

Billet achieved the necessary condition of coherence and small separation by a "split-lens" device shown in Fig. 4.4. The lens L is split into two and the halves slightly separated with a fine screw. A point source S produces two real images S_1, S_2 the

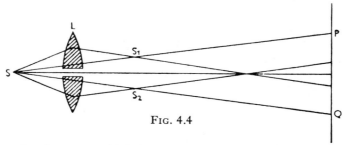

Fig. 4.4

separation between which is dependent on the separation of the halves of the lens. The two images S_1, S_2 are coherent sources and as a result interference fringes of the Fresnel type appear in the overlap region and can be received on a screen PQ.

Owing to the non-symmetrical character of the half-lenses the images are somewhat broadened, but this does not seriously affect definition.

An interesting modification of the Billet split-lens is due to Meslin (1893) who displaced the two halves of the lens *along* the axis as shown in Fig. 4.5.

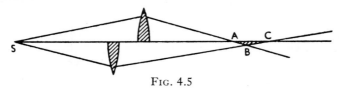

Fig. 4.5

Two images are formed from the point source S. Hence we have the case of two sources *behind* each other in line and as shown in Chapter 1 this should lead to *ring*-fringes on a screen. However, the two beams only overlap in the region ABC, which is that covered by both. The ultimate effect is the production of *half*-rings in this region, which can be regarded as a restricted region of non-localization.

Fresnel's Three-mirror Experiment

It had been known since the time of Newton that the central fringe in Newton's rings is dark, and Young concluded from this that when light is reflected from a rare medium at the boundary of a dense medium there is a phase change of π, i.e. an apparent alteration in path of $\lambda/2$. To confirm this suggestion Fresnel, in 1819, devised his three-mirror experiment which is illustrated in Fig. 4.6.

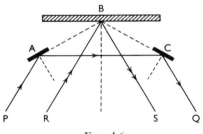

Fig. 4.6

A light beam P is reflected from mirror A to mirror C and then to Q. A parallel ray R is reflected from mirror B to S. By arranging the angles so that the grazing angle from R to B is twice the angle between A and B this condition is achieved and the rays S and Q have the same geometrical path difference and are in a condition to interfere. Fresnel found that with such an arrangement the central fringe is *dark*. Since one beam has been twice reflected, this observation confirmed that a phase change of π takes place on reflection.

Lloyd's Mirror Experiment

Whilst Fresnel required *three* mirrors to show this, a simpler arrangement which achieves the same result with *one* mirror is due to Lloyd (1837) whose arrangement is shown in Fig. 4.7.

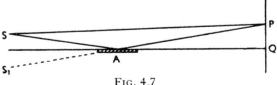

Fig. 4.7

Let A be a *front-surfaced* mirror, and let S be a source which is placed close to the plane containing A. Light reflected at A forms a virtual image S_1 which is close to and coherent with S. On a screen P, interference will take place between the two rays SP and S_1P and the fringe separation can be calculated as before in terms of the distances SS_1 and SP.

However, the conditions of observation differ from those of Fresnel's mirror experiments for there can only be light in the upper half of the field QP. If a thin piece of mica is interposed into the direct beam, the fringe pattern is displaced up above Q and the centre order (in this case the first fringe) can be seen. It is found by this means to be *dark*, proving again that there is a phase change on reflection.

Phase Change and the Reversibility Principle

By postulating a simple Principle of Reversibility, Stokes showed that a difference in phase must occur between reflection from say an air-to-glass interface as compared with the reflection at a glass-to-air interface. The Reversibility Principle simply states that, in any ray-tracing procedure, reversal of all the rays produce the original, if there is no absorption. Consider Fig. 4.8.

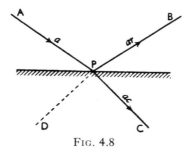

Fig. 4.8

Let an incident ray AP of amplitude a reflect to B, with amplitude ar and refract to C with amplitude ac. Now reverse these two rays. BP gives rise to a reflected ray ar^2 along PA and a refracted ray arc along PD. The ray CP reflects at P to a ray along PD, but it will not be assumed that the reflectivity for the direction glass → air is also r, instead let it be r', hence the reflected amplitude along PD is acr'. CP also produces a refracted ray along PA and let this be acc'. Since there was no ray PD initially then according to the Reversibility Principle it follows that:

$$acr' + arc = 0$$

Also for the PA direction:

$$ar^2 + acc' = a$$

Thus:

$$r' = -r \qquad . \quad . \quad . \quad . \quad . \quad (4.1)$$

and

$$r^2 + cc' = 1 \qquad . \quad . \quad . \quad . \quad (4.2)$$

The physical interpretation of the negative sign in (4.1) is that both reflectivities have the same numerical value, but a phase change relative to each other, amounting to π, as already assumed from observation.

The second relation, $r^2 + cc' = 1$ will not be required here.

The Visibility of Interference Fringes

In what has gone before an important simplification has been adopted, for it has throughout been assumed that both interfering sources have the same *intensity*. This is by no means always the case in interferometry, and we shall proceed to consider what happens if the two coherent sources differ in intensity. Such a condition could easily arise. Suppose, for example, in a Fresnel two-mirror experiment the reflectivities of the two mirrors differ. Or again consider the Fresnel biprism. In practice it is made of a single block of glass, but one could in theory make a biprism from two prisms with different absorptions to light, one glass being denser than the other, or perhaps containing an absorbent. Indeed in Lloyd's mirrors the two images have similar, yet different intensities, and when we come later to discuss interference by *division of amplitude* we shall find frequently that interference takes place between beams of unequal intensity. The following simple treatment is therefore of some importance.

In discussing the Principle of Superposition in Chapter 1 it was shown that the combination of two coherent waves of different intensity, $a_1 \sin \phi$ and $a_2 \sin (\phi + \theta)$, and with a path difference d, resulted in a wave with the same period but with a change in amplitude and phase, the resultant being

$$Y = \sqrt{\{(a_1^2 + a_2^2 + 2a_1a_2 \cos \theta)\}} \sin (\phi + \epsilon)$$

where $\qquad \theta = \dfrac{2\pi d}{\lambda} \quad$ and $\quad \tan \epsilon = \dfrac{a_2 \sin \theta}{a_1 + a_2 \cos \theta}$

The intensity is proportional to the square of the amplitude factor, i.e. to $(a_1^2+a_2^2+2a_1a_2 \cos \theta)$. As $\cos \theta$ varies from $+1$ to -1 the intensity varies from a maximum of $(a_1+a_2)^2$ to a minimum of $(a_1-a_2)^2$. Maxima occur when $d = n\lambda$, minima appear at $d = (n+\frac{1}{2})\lambda$.

The intensity takes on a sinusoidal form but does not go down to zero at the minima.

If I_{max}, I_{min} are the respective intensities of the maxima and minima, then the quantity $V = (I_{max}-I_{min})/(I_{max}+I_{min})$ (which was introduced by Michelson) is called the "visibility" of the fringes. It is a measure of the fringe clarity. The visibility can vary from a maximum of $+1$ (when I_{min} is zero) to a minimum of 0 (when $I_{min} = I_{max}$), in which latter case there are of course no fringes to be seen.

In practice owing to the difficulty of detecting intensity differences visually of less than 20% (the Rayleigh criterion) fringes frequently lose their identity long before zero visibility. If one assumes that it is only just possible to observe the minima when $I_{min} = 0\cdot8\,I_{max}$, the corresponding visibility is $0\cdot11$. This occurs for two sources with intensity ratios 360:1, which is surprisingly large. They must of course be coherent. Fringes of such low visibility have little value in practice.

It should be recognized that *visibility*, which is a quantity often emphasized in text-books, is frequently useless as a criterion of sensitivity. It will be shown later that fringes produced by multiple-beam methods are perhaps 100 times as sensitive as $\cos^2$-type fringes given by two beams, yet because the minima in the former do not go down to zero whilst they do in the latter, there exists the paradox that the multiple-beam sensitive, high-precision, easily seen fringes have a *lower* theoretical visibility than the less sensitive broad $\cos^2$-type fringes. It will become clear that the visibility criterion is not enough to define the practical sensitivity of fringe systems.

The Effect of Diffraction

In all the cases yet considered effects due to diffraction superposing its influence upon direct interference have been intentionally disregarded to retain simplicity. These effects are by no means inconsiderable, but their calculation is often a matter of some complexity. In the Fresnel two-mirror and Fresnel biprism experiments the common edge dividing the optical components acts as a cut-off and leads to characteristic diffraction effects, as a result of which there is an overall variation of intensity superposed upon the interference pattern leading to an effect which is quite marked.

In the Lloyd's mirror experiment, owing to the grazing incidence the effective aperture on viewing from P is so small that the mirror appears almost as a slit and all the associated diffraction variations of intensity consequent on viewing a narrow slit intervene.

However, these diffraction effects are mainly of secondary interest, but in one particular case of interference the diffraction is of some importance especially in connection with applications to instrument design. This particular case is that of the interference produced at the focus of a lens when normal incident light falls on to *two* narrow slits. One obtains then a compound effect due to Fraunhofer diffraction and Young interference. An approximate theory of this will now be discussed.

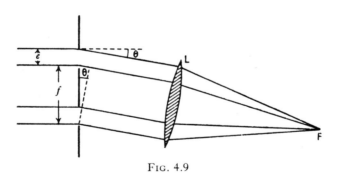

Fig. 4.9

Two-slits: Combined Interference and Diffraction

Let a plane wave at normal incidence fall upon two identical slits of width e and separated by a distance f between corresponding parts, as shown in Fig. 4.9; e and f are small compared with slit lengths.

If the diffracting beams in the direction θ are brought to a focus F by a lens L then at this point combined diffraction and interference takes place in accordance with the following. In the case considered θ is not large.

Considering first *a single slit*, if A is the amplitude emerging in the direction of the normal, then as shown earlier the amplitude in the direction θ is

$$a = (A \sin \alpha)/\alpha$$

in which

$$\alpha = (\pi e \sin \theta)/\lambda \doteq \pi e \theta/\lambda$$

Thus

$$a = \frac{A \sin \pi e \theta/\lambda}{\pi e \theta/\lambda}$$

Now the resultant amplitude produced by combining *two* light waves of the form

$$y_1 = a \sin \phi \quad \text{and} \quad y_2 = a \sin (\phi + \psi)$$

is

$$y = 2a \cos (\psi/2) \sin (\phi + \epsilon)$$

in which

$$\tan \epsilon = \sin \psi/(1 + \cos \psi)$$

The intensity, being the square of the amplitude term, is

$$I = 4a^2 \cos^2 (\psi/2)$$

in which ψ is the phase difference between the two beams and is clearly $(2\pi/\lambda)f\theta$ for corresponding beams from the two slits.

Thus

$$I = 4a^2 \cos^2 (\pi f \theta/\lambda)$$

Now putting in the value for a given above leads to

$$I = 4A^2 \cdot \frac{\sin^2 (\pi e \theta/\lambda)}{(\pi e \theta/\lambda)^2} \cdot \cos^2 (\pi f \theta/\lambda)$$

The intensity is then the product of a diffraction term and an interference term.

If f is much greater than e, then a given change in θ produces a far larger effect on the $\cos^2$ term than on the $\sin^2$ term. The $\cos^2$ term alone goes through maxima and minima, having maximum values of 1 when θ has the values θ_1 given by:

$$\pi f \theta_1/\lambda = 0, \quad \pi, \quad 2\pi, \quad 3\pi, \quad \text{etc.}$$

i.e. when

$$\theta_1 = 0, \quad \lambda/f, \quad 2\lambda/f, \quad 3\lambda/f, \quad \text{etc.}$$

and zero values half-way between.

This is shown in Fig. 4.10 (a) and represents the simple case of interference with two sources, i.e. a $\cos^2$ variation of intensity.

The other factor, taken by itself, gives the diffraction pattern of a slit, in which the maxima occur successively at (nearly)

$$\theta_2 = 0, \quad (3/2e)\lambda, \quad (5/2e)\lambda, \quad \text{etc.}$$

with minima at

$$\theta_2 = \lambda/e, \quad 2\lambda/e, \quad 3\lambda/e$$

As

$$e \ll f \quad \text{then} \quad \theta_2 \gg \theta_1$$

and the distribution is as shown in Fig. 4.10 (b).

The intensity I, being the product of these two distributions, has the final form shown in Fig. 4.10 (c).

This then shows the intensity variation of the interference fringes due to the superposition of the diffraction envelope.

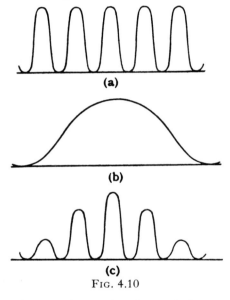

(a)

(b)

(c)

Fig. 4.10

In the next chapter the applications of the two-beam interference effects to the making of physical measurements will be considered. An instrument which makes use of interference fringes for the determination of some physical property is called an interferometer. Interferometric measurements have high precision largely because the unit of measurement is the light wavelength, a quantity of the order of only 5×10^{-5} cm.

Broadly speaking they are of two distinctive kinds, one in which the total light paths are *small*, the other in which the light paths are *long* but the path *differences* are still kept small.

CHAPTER 5

INTERFEROMETERS EMPLOYING DIVISION OF WAVE-FRONT

The Rayleigh Interferometer

The Rayleigh interferometer (1896) is a simple device enabling refractive indices of gases to be measured with precision by using interference from two slits as shown in plan in Fig. 5.1 (a).

Light from a slit source S is made parallel by a lens L_1, and the two wide slits S_1, S_2, about 1 cm apart, are illuminated at normal incidence. After passing respectively through the tubes T_1, T_2 and through glass plates G_1, G_2 the lens L_2 brings the

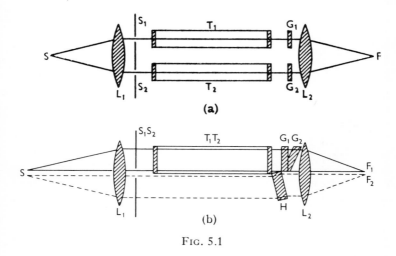

Fig. 5.1

beams to a focus at F where interference fringes are viewed with a cylindrical lens. Essentially they are like Young's fringes, with the overall modification in intensity due to diffraction.

If T_1 and T_2, both of length l, are evacuated, the fringes at F will be symmetrically situated with respect to the optic axis. If gas of refractive index μ is introduced into one tube, a path difference $(\mu-1)l$ between the two interfering rays is produced and the

43

fringes shift a number of orders n such that $n\lambda = (\mu - 1)l$. If monochromatic light is used, then n cannot be determined, hence a special fiduciary system of fringes (first used by Haber and Löwe) act as a comparison and these are formed in the way shown by the elevation diagram (Fig. 5.1 (b)).

The tubes T_1 and T_2 only cover half the slit length and produce fringes, F_1, above the half-way line. Light passes below this line and produces a fixed fringe system, F_2, which acts as a reference to which F_1 is compared. The inclined glass plate H is a simple

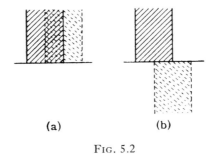

(a)   (b)

Fig. 5.2

device for bringing up the lower reference system to touch the upper system and avoid the obstruction due to the wall thickness of the tubes. The eye is sensitive to the detection of quite small displacements with such a fiduciary reference system, and the "resolving power" much exceeds the Rayleigh limit. This is made clear by considering the resolution of two broad fringes alongside, as in Fig. 5.2 (a), and displaced vertically, as in Fig. 5.2 (b). In Fig. 5.2 (a) the resolution is much less effective than in Fig. 5.2 (b). Experience proves that in Fig. 5.2 (b) a smaller separation can be detected than in Fig. 5.2 (a).

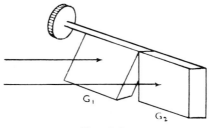

Fig. 5.3

In practice a null method is employed, use being made of the compensating device G_1G_2 which is shown in more detail in Fig. 5.3. This was introduced into interferometry by Jamin in 1856. The Jamin compensator consists of two identical plates of glass set at an angle on a bar which can be rotated by a graduated head. The two glass plates are in the paths of the two separate interfering beams. A rotation of the head about the horizontal axis introduces a slight differential path difference because of the different inclinations of the plates. The angle between these can be varied, the smaller the more sensitive is the set-up.

The compensator is first calibrated in terms of a standard wavelength λ, the number of divisions to move one fringe order relative to the fiduciary system being noted. The white-light fringes are used in the refractive index determination and the displacement of the central order is immediately recognized when gas is admitted. The two sets of fringes are brought into coincidence with the compensator, and from the required rotation of the compensator head the optical path length introduced is obtained, since the previous calibration in standard wavelength permits conversion of the compensation angle into direct wavelength retardations.

The Rayleigh Interferometer in Practice

The interferometer has been used both for gases and for liquids. For gases a tube length of 1 metre can be tolerated if the tubes are well lagged thermally. For liquids it is not practicable to employ path lengths exceeding 10 cm, since temperature gradients introduce considerable difficulty.

The fringes have a $\cos^2$ intensity distribution, hence light and dark spaces are of equal extension and as such are difficult to measure. Yet it has been claimed that a displacement of one-fortieth fringe can be evaluated. If a 1-metre path is used and $\lambda = 5 \times 10^{-5}$ cm, then a simple calculation gives the change in μ which can just be recognized as $1 \cdot 25 \times 10^{-8}$. The compensator, in practice, cannot introduce much more than 100 wavelengths and this restricts the range of μ that can be evaluated.

The apparatus has many uses. The gas type has been used to follow refractive-index changes in gases consequent upon chemical reactions, and the liquid type has industrial and biological applications in connection with refractive index determinations.

Because the two interferometer slits are of necessity wide apart in order to permit of sending separate beams down the two tubes, the fringe separations are minute and high magnification is necessary. Rayleigh therefore viewed the fringes at F with a short-focus cylindrical lens (a piece of selected glass rod). The magnification is required only in one direction and a considerable saving of light accrues.

Väisälä's Interferometer

An unusual modification of Young's slits has been introduced by Väisälä (1927) for the purpose of measuring very long paths (say 100 metres) for such purposes as geodetic survey, etc. Despite the long path length the usual precision of optical interference is achieved. The first arrangement used is shown in Fig. 5.4. A parallel beam of light from S falls at normal incidence upon two slits S_1, S_2. The light is reflected from the double-sided mirrors ABC and at F produces interference. When the path lengths AB and BC are equal white-light fringes are obtained even if AB = BC is very long. AB is a standard length and thus, by adjustment of C, BC can be made equal to this.

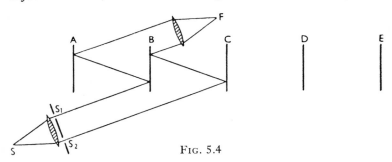

FIG. 5.4

The beams are then shifted so that BC, CD are compared and so on, stepping up to E, etc. Ultimately an extremely long path B—X is evaluated with interferometric precision in terms of AB. It is possible to measure paths of sufficient length for survey purposes, the limit being set by the irregularity of atmospheric conditions and earth tremors.

In a later modification Väisälä employed a step-up procedure in accordance with the arrangement shown in Fig. 5.5. The path length e_2 is n times the shorter length e_1, which is a standardized metre length measured against a substandard (in the illustration $n = 3$). The length e_2 can be adjusted. When white-light fringes are obtained the difference between e_2 and ne_1 is less than a light wave. If the difference exceeds a few wavelengths, no fringes are visible at all.

The step-up ratio (n) may be of some 5 or more, but increasing n too far leads to a serious reduction in visibility since the nth beam is reduced in intensity through the reflectivity R of the mirrors being less than unity. Thus if both incident beams are given intensity unity, that emerging from the e_2 range has intensity R^2 and that from the e_1 range, intensity R^{2n}. The intensity ratio of the two emergent beams is then $r = 1/R^{2(n-1)}$. Taking R as 80% (a possible value) say with aluminium mirrors, then for $n = 5$, r has the value 6, but for $n = 10$ it is 55, and with such a marked

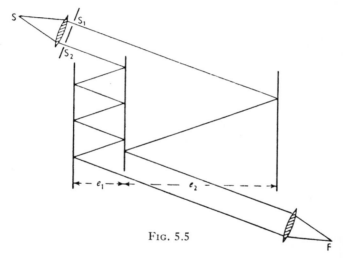

FIG. 5.5

intensity difference fringe visibility becomes very poor. The value of n is therefore restricted to small values. It can be shown that in a related type of interferometer employing partially silvered mirrors a step-up system with n as high as 16 is permissible.

In this Väisälä system once e_2 is evaluated the step-up procedure can be repeated to a new value $e_3 = ne_2$. However, there are a number of restrictions which limit the arrangement to but a few steps. Thus, excluding external disturbances, the longer paths involve opening up slit widths because of intensity difficulties, and definition suffers accordingly. Furthermore, each change in path requires an alteration of the angles of incidence and this introduces mechanical difficulties. Again, with long distances the *lateral* displacements of the beams down the mirrors is considerable unless the incidence is very nearly normal.

It is clear that the experimental procedure is one of some difficulty and the measurement of paths of the order of 100 metres requires the utmost care and skill, and much experimental ability.

Fizeau's Double-slit Experiment

Fizeau carried out an experiment in 1868 which was to have a most important application when developed twenty years later by Michelson. Up to the present, discussion has been concerned with fringes produced by dividing the wave-front emerging initially from a *single* source. Fizeau considered the appearance in the field of view if a *double* source was used, for example, two

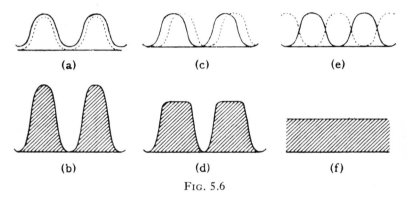

FIG. 5.6

stars close to each other, restricting observation to one wavelength. It is to be emphasized that since there is no coherence between two such sources *there can be no interference effect between them.* On the other hand it is possible to produce interference fringes independently from each, and by suitable arrangement the two systems can be superposed. If one system can be moved relative to the other, then the fringes can be maintained sharp, or made to disappear if the two sources are of similar intensity. If the two systems (Fig. 5.6 (a)) superpose, the result is clear bright fringes as in 5.6 (b). If on the other hand one system is displaced by about quarter of a fringe (5.6 (c) and 5.6 (d)), a broadened pattern results. For a complete half-fringe displacement (5.6 (e) and 5.6 (f)) illumination is uniform and fringes disappear. Thus the visibility depends on the fringe displacement and can vary from 1 to 0.

Suppose a distant source is viewed through a telescope which has two slits over the objective distant a apart as in Fig. 5.7, then $\cos^2$ fringes are formed at F and the angular separation between successive fringes is λ/a. Suppose now another source is viewed and let the two sources subtend an angle θ at the telescope, the $\cos^2$ fringes are produced at F_1, with the same separation, λ/a. If the F_1 fringes are displaced half an order, then the two fringe

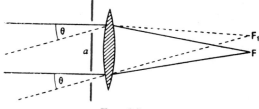

Fig. 5.7

systems will be in dissonance and this leads to zero visibility. This takes place when $\theta = \lambda/2a$ or $3\lambda/2a$, $5\lambda/2a$, etc.

Fizeau was by this means able to measure the angular separation of a double star by varying a until the fringes·disappeared. He considered whether the method could be used to measure the *diameter* of a star. All stars are so distant that even the largest available telescope shows no disc, other than the diffraction Airy disc. The observation on the double star (2 sources) can be

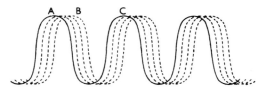

Fig. 5.8

extended to one continuous *line*-source according to the following treatment given by Williams.

Imagine a distant *line-source*, then, instead of two beams incident normally and at θ, there is a continuous range of angles of incidence between 0 and θ. As shown in Fig. 5.8 each point on the source has its own fringes (dotted) displaced from those at normal incidence (continuous), and the extent to which these encroach in the space between adjacent fringes of the normal beam depends on θ. The combined result shows that until the dotted region A to B reaches C, there will still be some fringe visibility. Thus θ can extend the *whole* angular fringe separation λ/a instead of half this value as was the case for two point sources. Taking into consideration the Airy disc, if the line is viewed by a circular lens, then the fringes will vanish if θ, the angle subtended by the line, equals $1\cdot22\lambda/a$ or multiples of this. The same considerations apply to the case of a continuous-disc source.

Application of Fizeau's Method

The first successful application of Fizeau's proposed method for measuring the diameters of celestial objects was made by Michelson in 1890. In 1874 Stephan had attempted by this method to measure the angular diameters of stars, but these are so small that his telescope had not a large enough aperture to permit of the slits being moved far enough apart to lead to dissonance in the fringes. Michelson therefore first examined four of Jupiter's moons, using the large Lick refracting telescope in California and found angular diameters of $1\cdot02 \rightarrow 1\cdot37''$ of arc. The experiment was carried out by means of symmetrically moving slits placed over the objective.

In 1899 Hamy suggested an important improvement, proposing to replace the slits by large rectangular openings, and showed that results could be obtained with increase in brilliance. This was later made use of by Michelson in his celebrated "stellar-interferometer" which will now be described.

Michelson's Stellar-interferometer

Since the aperture of the telescope used on Jupiter's satellites was insufficient to show the disappearance of fringes when turned on a star, Michelson (1920) used the ingenious device of effectively increasing the distance between the slits by building a girder system on to a telescope as shown in Fig. 5.9.

The large 100-in. Mount Wilson *reflecting* telescope was selected in order to be strong enough to hold the mount. The ray-tracing diagram is simplified but remains unaltered in essentials if we construct the conditions for a *refracting* telescope and this is done in Fig. 5.9. Four mirrors A, B, C and D are mounted as shown, B and C being fixed (separated 45 in.) and A and D movable such that they remain respectively parallel to B and C and equally separated from them.   Direct light from the star is prevented by a stop E from entering the telescope, only the reflected beams from A and D being admitted.   It is clear that A and D are equivalent to the slits in the Fizeau experiment, modified into apertures as suggested by Hamy.

The two mirrors B and C are intentionally small for then the effective aperture of the telescope is reduced and at F there is quite a large Airy diffraction disc.   (In the experiment the fringes at F are viewed with a microscope under high magnification (×1600)).   Under such conditions the Airy disc seen appears to be quite large and covered with fringes.   Had full aperture been

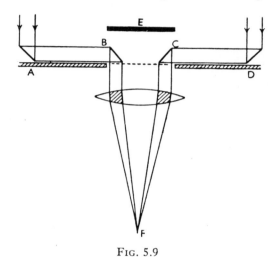

FIG. 5.9

used the image at F would have been very small.   The separation between the fringes is determined by the separation of the two mirrors A and D in accordance with the treatment already given. The magnification used is ample, despite the fineness of the fringes occasioned by the large distance.

The outer mirrors A and D are moved symmetrically until at a distance d the fringes vanish, then θ, the angular diameter of the star, is $1\cdot22\lambda/d$.

The first star examined was Betelgeuse in the constellation of Orion.   The fringes were found to vanish with $d = 121$ in.   Of course no fringes would be seen if the apparatus had gone out of adjustment during the moving of the mirrors hence the telescope was turned on to another star. The fringes thereupon reappeared, proving in a striking manner that the correct conditions had indeed been achieved for Betelgeuse.

Before θ can be calculated it was necessary to determine λ and this was done in two ways. In the first method the separation between the fringes x was measured, and as we have already seen if BC $= a$, then $x = \lambda f/a$, f being the focal length of the lens. Starlight is too weak to permit of monochromatization, thus either a coarse filter must be employed or the star must be viewed directly. In both cases there is a range of λ values, but a mean is obtained which suffices. An alternative method for arriving at a mean wavelength was for the observer to estimate the colour of the star within the spectral range and adopt the corresponding wavelength in the spectrum for that colour as the mean value. The colour seen depends upon the visual acuity of the observer, but so does the condition of disappearance of fringe visibility. Thus it is essential that both observations be carried out by the same experimenter. The value found for the angular diameter of Betelgeuse was 0·047 second of arc.

The Michelson stellar-interferometer is of great importance for, from stellar diameters, stellar densities can be determined, and fundamental theoretical arguments relating to the structures of stars depend upon these measurements. An extensive programme is being carried out in observatories. The resolving power available is very great, and there is no reason why girder lengths should not be extended at will, to be in fact some hundreds of feet across, with perhaps the possibility of determining angular diameters of the order of 0·001 second of arc.

It is of much interest to point out that the principle of Fizeau's double-slit method for angular resolution has been successfully applied at the other extreme of physical dimensions, i.e. to ultramicroscopic particles. By placing two slits before a microscope objective it is possible, in exactly the same way as was considered for the star, to obtain dissonance of fringes from an illuminated ultramicroscopic particle and thus to derive its diameter even though the diameter be far too small to be resolved by a microscope employed in the normal fashion.

CHAPTER 6

INTERFERENCE BY DIVISION OF AMPLITUDE: WEDGE-FRINGES

Introduction

In previous chapters consideration has been given to interference effects produced by the division of the wave-front into two separate parts which then combine. Of equal importance regarding practical application are the interference effects produced by division of *amplitude*. This division is usually brought about by simple reflection. Thus when light falls on to a thin film, say of glass or mica, part of the amplitude of the incident wave is reflected at the first face of the film met and the remainder continues until it meets the second face where a similar fraction is reflected. These two fractions being coherent and having traversed different path lengths are in a condition to interfere.

The first interference effect studied seriously (Newton's rings) involved this division of amplitude, but there are many other important applications. With interference between plane surfaces the effects produced depend upon whether the reflections (or transmissions) arise from a parallel-sided reflecting film of transparent material, or from a film in which the faces are inclined to each other, the latter case usually being called a wedge. In the former case the fringes occur at infinity, but the wedge-fringes are localized. It is intended to discuss wedge-fringes in detail first, but the calculation of the law governing the fringe separation is more simply deduced from the interference effect in the parallel-sided film. This law, which shall be called the "cosine law", is the fundamental equation which describes and relates a considerable number of interference phenomena.

The Cosine Law

Consider parallel rays (Fig. 6.1), A_1, A meeting the surface B_1B of a plane parallel sheet of material of thickness t and refractive index μ. The ray A_1B_1 refracts to C_1 where it reflects to B. At B it meets the ray AB and both travel towards C. The two rays are in a condition to interfere.

From B_1 and B drop respective perpendiculars B_1D and BD_1. The path difference in the air is BD. In the medium the path length is B_1C_1 plus C_1B. The time taken for B_1 to reach D_1 in the medium is the same as that for D to reach B in air. Thus the path difference Δ is $B_1C_1+C_1B-B_1D_1$ in the medium. Now produce B_1C_1 its own length to E and join BE. The path difference Δ equals D_1E. As BE $= 2t$ then $D_1E = 2t \cos \theta$ in which θ is the angle of refraction in the medium. In air the equivalent path difference is then $2\mu t \cos \theta$.

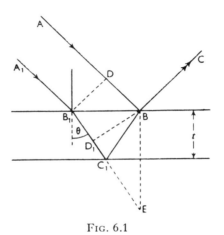

FIG. 6.1

Now it has already been shown that reflection at B will introduce a phase change of π which is equivalent to altering the path length by $\lambda/2$. Since the beams would have reinforced when the retardation is an integral number of waves (if not for the phase change), it follows that when $n\lambda = 2\mu t \cos \theta$ we get interference producing a *dark* fringe. This is the cosine law. Bright fringes form for $(n+\frac{1}{2})\lambda = 2\mu t \cos \theta$.

Classification of Fringes

The basic formula $n\lambda = 2\mu t \cos \theta$ contains the possible variables μ, λ, t and θ which determine n, the order of interference. In general, μ cannot easily be varied and in very many experiments it is constant; often it can be taken as unity. This leaves λ, t and θ as the frequent variables, and on this basis a classification of fringes can be made which will prove helpful. Here we consider only the case when λ is constant, as follows:

Nature of light	Constant	Variable	Name of fringe
λ constant (monochromatic light)	θ	t	equal thickness
	t	θ	equal inclination

Nature of Wedge-fringes

If a wedge of quite small angle is considered, then the basic formula $n\lambda = 2\mu t \cos \theta$ derived for a parallel-sided film will still hold closely enough. With such a system the fringes will appear as in Fig. 6.2, at those positions A, B, C, . . . on the upper surface at which the wedge thicknesses t are $\lambda/2$, $2\lambda/2$, $3\lambda/2$, . . ., etc., if for convenience it is assumed that the incidence is normal and μ is unity.   If the wedge is of a dispersive material then it is μt that equals $\lambda/2$, $2\lambda/2$, . . ., etc.

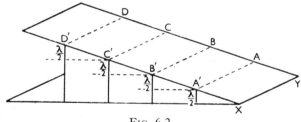

Fig. 6.2

A line such as AA′ is the locus of points for which n is constant, since, for such a line parallel to the edge of the wedge XY, t is constant. Thus straight-line fringes are formed and these are called "fringes of equal thickness", for although t varies on moving from Y to D, yet each fringe is a traverse of points for which the wedge thickness is constant. The fringes can be considered as height-contour lines exactly as on a geographical contour map. If either of the two surfaces is irregular, then the fringes will contour the irregularity, each following a path of constant optical thickness of the air film.

The Localization of Air Wedge-fringes

It is often stated that wedge-fringes are localized in the wedge, but this is only true at normal incidence. The calculation of where the fringes are sharply localized in the general case is complex but a simple derivation can be used for an air wedge, i.e. a wedge in air between two plane glass plates. In this case it is permissible to disregard refraction effects since the plates do not alter the *directions* of rays which emerge from them. A simple treatment can be given as follows: Let PQ, RS be the wedge, the angle ϕ being quite small (it is schematically enlarged in Fig. 6.3). Let a ray AB, incident at angle i reflect a

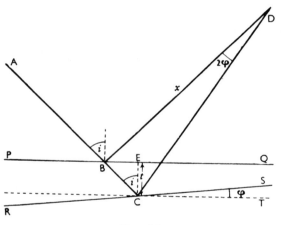

FIG. 6.3

fraction to D and send the remainder to C. At C the beam reflects a similar fraction to that of B and ultimately CD meets BD at D. Because SCT $= \phi$, then angle BDC is 2ϕ. The wedge thickness changes only quite slowly, and if the thickness at C is EC $= t$, then BC $= t/\cos i$. Now angle BCD is very closely $2i$, since ϕ is small, hence in triangle BCD, writing BD $= x$, then

$$\text{BD}/\sin 2i = \text{BC}/\sin 2\phi = \text{BC}/2\phi \quad \text{for } \phi \text{ is small}$$
$$\therefore \quad x = (t/2\phi \cos i)2 \sin i \cos i$$
$$= (t/\phi)\sin i.$$

This quantity does not involve the wavelength of the light. Since the two rays meet at D the interference fringe is located at this point and the quantity x gives the distance of this one fringe from the point of incidence on the front surface. It can be shown that the path difference of the two rays on reaching D is $2t \cos i$ (closely enough).

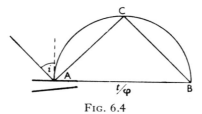

Fig. 6.4

To appreciate practical values, consider light incident at an angle of 30° on a thin film at a section where t is $\frac{1}{10}$ mm thick. Let the wedge angle be such that fringes of wavelength 5×10^{-5} cm are 1 mm apart when there is normal incidence (a typical case).

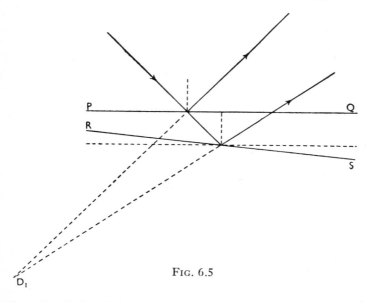

Fig. 6.5

Then $\phi = 2 \cdot 5 \times 10^{-4}$. Substitution gives $x = 20$ cm for the 30° incidence. Thus fringes can be localized at quite appreciable distances from the film producing the interference effect.

The following construction, Fig. 6.4, is suggested to show how the localization varies with angle of incidence i. Let A be the

wedge-film. In the plane of the upper surface lay off a length
AB $= t/\phi$. On this draw a semicircle, then this semicircle is the
locus of the points of localization on which the fringe will be
found. The proof is as follows: For a ray incident at angle i the
reflected ray is AC in which angle CAB is $(\pi/2)-i$. As ACB is
a right-angled triangle in which AB $= t/\phi$, then the length AC $=$

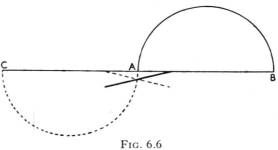

FIG. 6.6

(t/ϕ)cos CAB $= (t/\phi)$sin i. Hence C is the localization of this
fringe, as already shown. This holds for all chords, corresponding
to all values of i. The fringe is only localized in the actual
plane of the film for $i = 0$ (normal incidence) and for $i = \pi/2$
(grazing incidence). [Note: the approximation is invalid at $\pi/2$.]

Referring back to Fig. 6.3 the construction showing the point

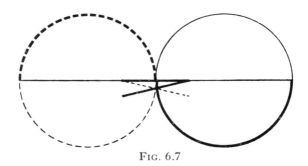

FIG. 6.7

of intersection D depends upon the fact that the wedge narrows
to the right, i.e. QS $<$ PR. If the wedge is in the opposite
sense, as in Fig. 6.5, then it is clear that a virtual image appears
on the lower side at D_1. From this it follows that the complete
localization curve for reflected rays with any arbitrary wedge
angle is formed by the two semicircles shown in Fig. 6.6, A being

the wedge and CB $= 2t/\phi$.   In the upper half the fringes are real and in the lower, dotted, they are virtual.

If now one considers the transmitted rays then, as it is easy to see, the completed diagram for localization for a given fringe is that in Fig. 6.7 in which the transmitted systems are shown in heavier lines.

The curves in Fig. 6.7 refer to the localization of a *single* fringe and it is necessary to determine the focal region of a succession of fringes.   Referring to Fig. 6.8, the path difference at D is $2t \cos i$. It is clear that a fringe B succeeds a fringe A when the air-path length diminishes by $\lambda/2$.   Thus the nth fringe away from A has a path difference $(2t-n\lambda)\cos i$ appearing for a wedge thickness

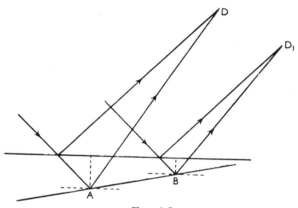

FIG. 6.8

$t-n\lambda/2$.   The radius of the circle of localization for the nth fringe will thus be $(t/\phi)-n\lambda/2\phi$.   However, this circle must be laid off from B_n instead of A.   Since the difference in thickness between A and B is $\lambda/2$ then it follows that the fringe separation s, between AB, is given by $s = \lambda/2\phi$.

Thus the circle from B_n is laid off from a distance $n\lambda/2\phi$ to the right of A.   Taking A as the origin, the far point (maximum distance) of localization for *all* circles must be

$$(t/\phi)-(n\lambda/2\phi)+(n\lambda/2\phi) = t/\phi$$

Thus all the semicircles meet and the construction is as shown in Fig. 6.9, A, B, C, D and E being the fringe positions on the wedge at normal incidence.

To find the localization for the whole fringe system for any given incidence i, draw the chords from ABCD . . . all parallel, making angle i with the normal. The line joining the chord ends is clearly the localization for the successive orders at the selected angle of incidence. Such a line of localization is shown as PQ. As i increases the fringes get closer together and farther away. From the geometry it is thus clear that the locus of localization is the intercept on the circles made by the line making angle i with the diameter. Thus by drawing such lines the locus of localization can be obtained for any angle of incidence. It is

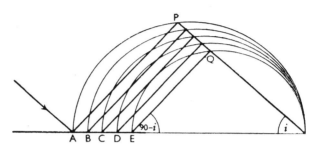

FIG. 6.9

clear too that the separation between successive fringes will vary since the intercepts between circles are $(\lambda/2\phi)\cos i$, and therefore vary from $s = \lambda/2\phi$ at normal incidence to zero at grazing incidence. In constructing the semicircle of radius $t/2\phi$ it is perhaps simpler to write this $s.t/\lambda$. Thus the radius is the product of the fringe separation times the number of waves in the thickness of the wedge.

The above somewhat simplified treatment is ample enough for the requirements of this treatise. It is, of course, only an approximation.

Localization of Wedge-fringes in a Dispersive Film

The calculation for the general case for any angle of incidence on a wedge consisting of a dispersive medium, e.g. glass, is complex and was first made by Feussner (1880), the surface of localization of wedge-fringes being thus often called the Feussner surface. The expression for the distance of the localized fringes from the point of incidence on the first interference surface can be shown to be

$$x = \frac{t}{\phi} \frac{\sin i \cdot \cos^2 i}{\mu^2 - \sin^2 i}$$

(provided one considers the simplifying case of incidence in a plane at right angles to the edge of the wedge).

For an air wedge ($\mu = 1$) this reduces to the simple expression already derived. But for glass, the expression shows that x is zero for both normal incidence ($\sin i = 0$) and for grazing incidence ($\cos i = 0$). The value of x is a maximum at approxi-

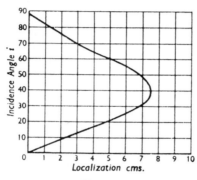

FIG. 6.10

mately $0 \cdot 2t/\phi$ for an average glass and this occurs at an incidence of about 40°. At this incidence the fringes for corresponding air wedges are about three times as far away.

A plot of x against i leads to a curve of lemniscate form as in Fig. 6.10 which was that found experimentally with a glass wedge $\mu = 1 \cdot 5$, $\phi = 1\frac{1}{2}'$ and $t = 1 \cdot 9$ mm.

In practice this localization of fringes with non-normal incidence is a matter of interest since it is met with in some experiments. This subject has been discussed here in some detail since it is ignored in many standard books on optics.

The Image Formation of Localized Fringes

By far the most important case in wedge interferometry is that of localization with normal incidence, in which the fringes appear on the surface of the film nearest to the eye (or viewing lens). Thus in reflection the fringes appear on the front surface, in transmission, on the rear surface. In the majority of cases the two surfaces are so close together that the fringe position can effectively be considered within the film.   Gehrcke (1905) has given a treatment which shows clearly the mechanism of projection by a lens of the reflection fringes localized in a film through the use of normal incidence with parallel light.   This treatment will now be discussed, but it is diagrammatically much clearer to represent the conditions in the case of *transmission*, although a strictly analogous construction holds also for reflection.   Transmission only is therefore treated below in Fig. 6.11 and it will be assumed that the rays interfering have similar intensities.   The figure shows a wedge WX, YZ of small angle ϕ illuminated by parallel light at normal incidence on WX.   The lens L projects fringes on to the screen FG according to the ray traces shown.

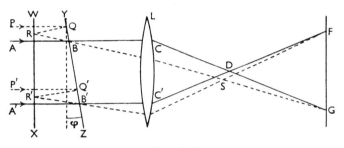

FIG. 6.11

Let a beam be defined by two extreme rays ABC and A'B'C', an image of the *source* being formed at D in the lens focal plane. Now interference fringes on the surface at BB' arise thus: A ray PQ (still normal incidence) strikes the surface YZ at a position Q such that after reflection at Q and successively at R it meets B and interference with the ray AB takes place at B.   Identically at B' the rays A'B' and P'Q' interfere.   However, the beam limited by the extreme rays PQ, P'Q' also form an image S of the *source* at the lens focal plane.

B and G are conjugate, thus an image of a *fringe* appears at G, likewise at F appears the image of the other fringe from B'.   The fringe separation BB' (or its image FG) depends on the value of ϕ for the increase in path thickness of the wedge corresponding to the two fringes must be $\lambda/2$.

In the above treatment the lens L is considered simply as projecting an image of the fringes, but an alternative view can be adopted.   It is legitimate to assume that D and S are two coherent sources quite close together directly producing fringes on the screen at the points F and G.   This treatment can be quite fruit-ful and has been used in discussing properties of multiple-beam fringes.

Complementary Systems

From energy considerations the transmitted and reflected fringes should be complementary and this is easily verified by experiment.   If a thin film is illuminated from both sides by sources of the same intensity, no fringes are seen either in trans-mission or reflection.   It should be noted that in reflection a phase change takes place because one beam reflects from a rare to a dense medium and vice versa for the other.   In transmission the reflected beam reflects under the same conditions twice, and thus a double phase change each of π brings the phase back to its original value.   For this reason the fringe systems are com-plementary, for clearly the two systems, transmitted and reflected, are displaced relative to each other by half a fringe by virtue of the phase-change effect.   Thus bright transmission fringes corre-spond to dark reflection fringes and vice versa.

Varieties of Thin Films

Interference can be observed in thin films which may be solid, liquid or gaseous (including vacua).   Low orders of interference can be studied with thin films of mica, or of glass, formed easily by heating the closed end of a glass tube and blowing out hard. It has been reported that glass films which are only a small fraction of a light wave thick can be blown by using a double capillary tube.   On heating and blowing the combination, the inner film so formed can be obtained exceedingly thin.

Thin films of liquid or gas can be studied by bringing together optically worked glass surfaces with the medium between them. The glass surfaces can be plane, or else slightly lenticular.   Some liquids can be brought freely into thin films, such as soap bubbles or foams.   Plane soap films can be drawn from soap solutions by wire frames, and the thicknesses of such films have been measured interferometrically.

Many liquid (and some solid) films can be made by placing a drop of a spreading liquid (or solution) either on a liquid or solid surface.   The spreading of oils over wet surfaces, or the formation of thin collodion skins by spreading collodion solution over water, with immediate rapid evaporation of the solvent are typical examples. Some films of interest are formed chemically, such as for example the thin coloured oxide films produced on metal surfaces when they are heated, copper and steel showing very characteristic films for example, the anodic oxidation of aluminium leads to films of controllable thickness.

Interference methods can be used to study the properties of such films, but perhaps of more significance is the converse application of the use of thin-film interference in many optical and metrological measurements. It should be noted that many experiments are best carried out with thin films, although thick films do also show interference. There are several reasons for this. Thin films imply low order of interference, hence white-light fringes can often be seen. High-order white-light fringes are not seen for the same reason as the inability to see more than a few low-order coloured fringes in say the Rayleigh refractometer, the higher-order coloured fringes superpose and obliterate visibility.

Of more importance is the fact that, in general, the two beams which reach the eye must come from a small surface area if either incident parallel or divergent light is used. This condition obtains if the two reflecting surfaces are close together. It is, however, possible to see fringes when surfaces are fairly widely separated if a pin-hole is placed before the eye to restrict the angle of the beam accepted for vision.

The following sections will treat of some of the more important applications of two-beam thin-film interferometry.

Newton's Rings

Described by Hooke, but named after Newton because of his analysis of their properties, Newton's rings have played so important a part in the history of optics that a considerable literature has been devoted to their discussion and use. They can be seen easily by merely placing a long focal length lens on a glass plate and viewing reflected light. As Newton described, with white light a few coloured rings appear, with colours appearing in a particular sequence, but with light even crudely filtered to one approximate colour, a considerable extension in the number of rings can be seen. Owing to the very close approach of the lens to the plate the air film between the two is quite thin and a consequence of this is that rings can be seen with a diffuse source, i.e. with one leading to a wide angular range of incidence. As Fizeau first indicated the definition is much improved by collimating, i.e. by using parallel light, the set up being as in Fig. 6.12 in which a source A (sodium flame) is at the focus of a lens B to give an incident parallel beam on the lens-plate combination L-P.

Light is reflected at the two faces enclosing the air boundary, with the formation of fringes localized in the air film and these can be viewed either by the eye, a hand lens or a low-powered microscope. If the two glass surfaces are pressed into near contact the central region, at which the path difference approaches zero, appears dark. Since there is effectively no path difference here one might have expected reinforcement of light, i.e. a central

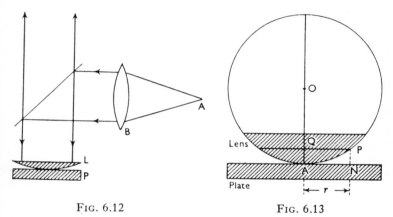

Fig. 6.12   Fig. 6.13

bright region. Thomas Young was the first to point out, at the beginning of the nineteenth century, that this dark region was evidence for the phase-change on reflection, for at the curved surface light is reflected at a glass-air interface, whilst at the plane glass surface at an air-glass boundary. The phase difference π leads to destructive interference for zero or near-zero path difference.

The geometry of the rings can be derived from Fig. 6.13.

With centre O complete the circle of radius R of which the curved lens surface forms an arc.

On moving out from the centre of the system A, let N be the position of the nth dark ring such that the path difference $2PN = 2t$ equals $n\lambda$.

But PN = QA, and if the radius of the nth ring, AN, is called r, then $r^2 = t(2R-t)$, or since QA is small compared with R:

$$r^2 = 2tR = nR\lambda \qquad . \qquad . \qquad . \qquad (6.1)$$

Hence $$r = \sqrt{n}.\sqrt{R\lambda}$$

From which it follows that the radii of the rings are proportional to the roots of the natural sequence of numbers $\sqrt{n}$.

If a liquid of refractive index μ be interposed between the lens-plate, the equation (6.1) becomes $r^2 = nR\lambda/\mu$. Thus the rings shrink in the ratio $\sqrt{(1/\mu)}$. Further, if the light is not incident normally, i.e. if we view at an angle θ the rings become ellipses, the major axes retaining the values r whilst the minor axes become $r_1^2 = nR\lambda/\mu \cos \theta$, a law confirmed by Newton himself.

In practice there is no certainty that the lens and plate are in true contact, and in any case to establish contact it might be necessary to deform the surfaces. This difficulty in computation can readily be overcome.

Suppose the pole of lens is distant x from the flat, then replace t by $x+t$ such that $2(x+t) = n\lambda$. For the mth ring $r_m^2 = R(m\lambda-2x)$, and for the nth ring $r_n^2 = R(n\lambda-2x)$. Subtracting gives $r_m^2-r_n^2 = R\lambda(m-n)$.

This eliminates x, and it is only necessary to measure the radii of any two rings, a known number of rings apart $(m-n)$.

It is self-evident that Newton's rings in transmission appear in complementary positions to those in reflection, but there is now the added complication that the two beams producing the fringes have very different intensities. The direct beam interferes with one which has suffered two reflections and are thus in the ratio of intensity of about 600:1. It will be recalled that the Rayleigh visibility limit is at a ratio of about 360:1, thus at strictly normal incidence the fringes are just below visibility. For incidence other than normal the intensity ratio becomes more favourable and faint fringes can be detected by a keen eye. The transmitted system formed by two glass surfaces can in general usually be disregarded.

Applications of Newton's Rings

Historically this was the first way in which the wavelength of light was determined. Newton measured λ from the rings, although he did not call the distance he measured the wavelength, but in accordance with his corpuscular theory, the distance between easy fits of reflection.

The classic determination by Fizeau in 1864 of the thermal expansion of crystals is an example of the use of Newton's rings for measuring small linear displacements and has served as an inspiration for many other measure-

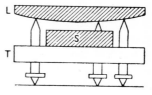

ments. Fizeau's dilatometric method is illustrated in Fig. 6.14. T is a tripod of platino-iridium and the three supporting screws hold on them a plano-convex lens L the curved surface having a long radius of curvature. The specimen S (a crystal or other object) is polished plane parallel, and Newton's

Fig. 6.14

rings are formed between the lower face of the lens, which has lines scratched on it, and the upper face of the object. Sodium light at normal incidence is used.

The whole is placed in a furnace, and the fringe displacement (relative to the scratched lines) on heating gives the difference in expansion between the object and the supporting tripod. The expansion of the tripod legs is separately evaluated by removing the object and bringing L down very close to T to form fringes between L and T.

It was with such an arrangement that Fizeau found that a change in gap corresponding to some 500 fringes led to a considerable reduction in fringe visibility, with visibility being re-established as good again when 1000 fringes had passed. He explained this as being a consequence of the fact that yellow sodium light consists of two wavelengths λ5890 and λ5896, i.e. lines with a separation of 6 Å which is almost 1/1000 of the wavelength of each. Thus when the path is about 500 fringes the maxima of the one fall on the minima of the other, there is partial obliteration and fringe visibility is poor. At 1000 fringes, concordance is re-established, but with one wavelength being exactly an order behind the other.

A considerable number of variants of the Fizeau dilatometer have been developed, and the tendency has been for complexities to be introduced. A simple type was developed at the U.S.A. Bureau of Standards shown in Fig. 6.15. Instead of using Newton's

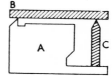

Fig. 6.15

rings, the lens is replaced by a glass plane and straight wedge-fringes are used.

The support A is a block of fused quartz at one end of which rests a plate B. The other end of B is supported by the specimen C, the size of which is carefully ground such that there is a suitable thin air film between A and B. The top of A is a polished plane. On heating, the relative expansion of C to A is determined from fringe movement in the film between A and B. Correction is made for the expansion of A by replacing C with an object of known expansion.

Fizeau Fringes

A most important step was taken for the whole subject of interferometry in the realization by Fizeau, in 1862, of the practical importance of strict collimation. Both from the basic equation $n\lambda = 2\mu t \cos \theta$ and also from a consideration of the conditions in Fig. 6.11, it is clear that precise sharp localization of fringes on a wedge depends upon using strictly parallel light. Now it is important to bear in mind the distinction between point and extended sources in this connection. Strictly speaking "parallel" light is only produced from a point source at the focus of a perfectly corrected lens, and of course this is a condition that can be approximated to. However, if one places an *extended* source at the focal plane of a lens one can conceive of the diverging light which emerges as consisting in fact of bundles of parallel light, each bundle coming from a different point on the source and apparently making different angles with the axis of the lens. One can describe this system loosely as producing "parallel" light but with a range of angles of incidence.

If such a system is used to illuminate a thin film, then beams of different angles of incidence lead to confusion thus: In the expression $n\lambda = 2\mu t \cos \theta$ this has maximum values at normal incidence ($\cos \theta = 1$). For light incident at any angle θ a given order n appears at a position in the wedge for which t increases. Thus if a range of angles of incidence is employed, all bright fringes broaden out *but on one side only*, i.e. the side corresponding to greater wedge thickness. It is clear from Fig. 6.11, even without mathematical analysis, that the amount of confusion depends upon the thickness of the wedge. For very thin films the disturbance is slight, for the linear displacement of beams along the surface is very small for even large angles of incidence.

If strictly collimated light is employed then it is clear that one is no longer restricted to thin films and indeed separations of surfaces of many *centimetres* can still produce good fringe definition. The basic idea of the Fizeau system of fringes is then the use of a small pin-hole at the focus of a good lens, and as the emerging light is effectively parallel, a viewing lens is needed. This is perhaps most easily seen if one disregards for the moment intensity differences in beams, and considers the system in *transmission*. The Fizeau arrangement would then be as in Fig. 6.16,

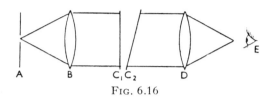

FIG. 6.16

in which A is an illuminated pin-hole (monochromatic light), B a well-corrected lens, C_1, C_2 the two surfaces, D a viewing lens, E the eye. The stricter the collimation the farther apart can C_1, C_2 be tolerated. The *angle* between these, *not the distance*, determines the number of fringes in the field of view, and if C_1, C_2 are parallel, then a uniform field of illumination is seen. If C_1, C_2 are separated, still parallel, the field of illumination goes through from light to dark as successive additions of separation increase by $\lambda/2$. For light economy it is best if the two lenses B and D are similar.

Now of course the above idealized system fails if C_1, C_2 are glass surfaces because of intensity differences. On the other hand, in reflection, perfect fringe visibility results. Thus in 1883 Laurent adapted the Fizeau system in reflection for revealing the the perfection (or imperfection) of optically worked surfaces and this remains to-day as a basic procedure in optical workshops.

Fig. 6.17 illustrates two variations of the arrangement of Laurent. In Fig. 6.17 (a) the illuminated pin-hole A is at the focus of the good lens L, and thus a parallel beam falls at normal incidence on to the wedge C_1, C_2. The light returns and, by the sheet of glass at 45° at D, fringes are brought to the eye at E. The system is economical since only one lens is used. If C_1, C_2 are not too far apart, a variation, Fig. 6.17 (b), is to displace A sideways and bring the light to the eye with a small prism P. This gives far more light, at the expense of non-normal illumination. For the glass plate D is a relatively poor reflector.

Since the system gives fringes of equal thickness, it is used extensively for mapping optical components. It can only be used for surfaces which are very nearly plane and the precision is set by the $\cos^2$ distribution of intensity. It will be shown later that modifications have been developed which lead to great improvements in sensitivity.

It has been claimed that the Fizeau fringes arise through simultaneous division of wave-front and amplitude. This is

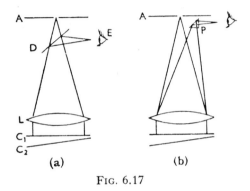

Fig. 6.17

certainly not the case. There is no *division* of the wave-front. There is division of *amplitude* and *displacement* of the two components thus provided. Indeed, in many cases of interference by division amplitude, there is also simultaneous *displacement* of the wave-front, but this cannot be classed as *division of the wave-front* in the sense already used to describe Young's and related fringes. Division of wave-front implies the existence of slits or apertures *side by side*, being met simultaneously by a wave, either spherical or plane, from a point or line source, and this condition is not at all fulfilled in the Fizeau arrangement.

Interferometric Reduction and Augmentation of Reflectivity

(a) *Blooming.* An important application of thin-film interference has been the recent development of techniques whereby the reflectivity of glass surfaces can either be diminished or augmented by the deposition on these surfaces of suitable thin interference films. The technique for reduction of reflectivity will first be considered. Since in the final product, the glass surface has a faintly purple colour, something like the bloom on fresh plums, the method has industrially come to be called "blooming".

It can be established that the reflectivity of a non-metallic surface depends upon both the refractive index of the material and that of the medium in which it is immersed. It depends also upon the angle of incidence and upon the state of polarization of the light, but if we restrict ourselves to normal· incidence the conditions are much simplified and polarization effects no longer exist. It is found that at normal incidence, for a glass of refractive index μ_2 in a medium of refractive index μ_1 the reflectivity is given by $[(\mu_2-\mu_1)/(\mu_2+\mu_1)]^2$. For a light crown glass ($\mu_2 = 1\cdot5$) in air ($\mu_1 = 1$) this has the value $(0\cdot5/2\cdot5)^2 = 0\cdot04$, whence it follows that the normal reflectivity of glass is some 4%, a figure well known from experience. For a dense flint glass ($\mu = 1\cdot66$) the reflectivity is 6%.

Now in complex optical systems, such as a camera lens, gunsight or periscope, there can be so many interfaces that as much as 75% of the incident light is reflected back and this has two serious defects. First, it reduces the light-gathering power of the system to a quarter of what it might be; second, the 75% reflected light is scattered and reflected at curved surfaces and can lead to bad haze effects, reducing visibility and definition. As long ago as 1892 Taylor discovered that tarnish films on lead glasses, produced by the action of H_2S, actually *increased* the transmission of light and we shall see now how this has been applied by Blodgett and Langmuir and by Strong (1935) to the controlled reduction of light reflectivity. In the modern techniques, a film of either magnesium fluoride or else sodium aluminium fluoride (cryolite) is deposited on the glass surface by thermal evaporation. The operation is conducted in a vacuum which should be at least as good as 10^{-5} mm Hg. The fluoride is heated from a tungsten or molybdenum filament and deposits as a very transparent uniform "glass" on the surface. Deposition is carried through *till the optical film thickness (i.e. the optical length) is one-quarter of a light wave*, for some specified wavelength, say $\lambda5000$. The theory of anti-reflection is then as follows. The essential feature is that the refractive index of the film of fluoride ($\mu_2 = 1\cdot36$) lies between that of air ($\mu_1 = 1$) and glass (μ_3, say $1\cdot5$). Consider illumination by monochromatic light of wavelength λ.

Since the film has $\lambda/4$ equivalent thicknesss then the optical path difference is $\lambda/2$ for in both instances the light meets a boundary by approaching from a region of lower refractive index and thus phase effects are the same at both boundaries.

Since the two reflected beams have path difference $\lambda/2$, they mutually destructively interfere.

Clearly if it can be arranged so that both beams have the same intensity then *all* reflection back will be totally annihilated.

At the air-film interface the fraction reflected (the reflection coefficient) R_1 is $\{(\mu_2-\mu_1)/(\mu_2+\mu_1)\}^2$ whilst at the film-glass interface the reflectivity R_2 is $\{(\mu_3-\mu_2)/(\mu_3+\mu_2)\}^2$. Now strictly speaking the beam which meets the film-glass interface is not quite as strong as that meeting the air-film interface for there has been reflection at the first face. However, since for $\mu = 1\cdot36$ the reflection at the first interface is only $2\cdot3\%$, the beam meeting the second interface is $97\cdot7\%$ of its initial value, and very little error will be introduced by assuming that the incident beams at both interfaces have the same value.

Clearly then for complete destructive interference we should have $R_1 = R_2$, hence:

$$\frac{\mu_2-\mu_1}{\mu_2+\mu_1} = \frac{\mu_3-\mu_2}{\mu_3+\mu_2}$$

On cross multiplying this immediately reduces to $\mu_2{}^2 = \mu_1\mu_3$ so that $\mu_2 = \sqrt{\mu_1\mu_3}$. Thus the condition requires that the refractive index of the film μ_2 should equal the square root of the refractive index of glass (for $\mu_1 = 1$). For light crown glass, with $\mu_3 = 1\cdot5$, μ_2 should be $1\cdot22$, and for a dense flint glass, with $\mu_3 = 1\cdot66$, μ_2 should be $1\cdot29$. Cryolite has a value $\mu_2 = 1\cdot36$ and this is a reasonably close approach. With it, crown glass reflectivity reduces from 4% to $1\cdot3\%$ and flint glass still better from 6% to $0\cdot6\%$. The abrasion properties of the film are important and as yet nothing better than the fluorides has been found.

Now it must be appreciated that there is in fact no loss in energy through the interference, merely a redistribution. What has been lost in reflection appears in transmission, and thus there is a double gain in that no light is lost and no scatter troubles ensue. Whilst the annihilation in reflection is an obvious mechanism, it may seem obscure as to why there is a simultaneous gain in transmission. Perhaps the easiest physical model to explain this is to imagine replacing each interface reflection by a source of light. In that case light travels both backwards and forwards and only destructively interferes backwards, whilst reinforcing forwards. This is, of course, merely a model for explaining the gain in the forward direction.

The above simplified theory applies strictly only to one wavelength, for clearly a given metrical thickness of film is $\lambda/4$ for one wavelength only. Thus over the spectral range of visible light the efficiency of the anti-reflector falls off on either side of λ. But the effect is not too serious. Magnesium fluoride, with a minimum reflectivity of $1\cdot3\%$ on crown glass at $\lambda5000$ has a reflectivity which rises smoothly to about $1\cdot8\%$ at both the red and violet ends of the spectrum.

The preparation of a $\lambda/4$ film is not difficult. It is only necessary to observe the reflection from the glass surface receiving the deposit during the deposition process, using a photocell to record the light reflected. This falls to a minimum when a $\lambda/4$ film is deposited and the coating action is then stopped.

It has been found possible by depositing two films of different refractive index to reduce the reflectivity to zero for one specific wavelength.

(b) *High-reflecting multilayers.* A converse procedure, also now of much technical importance, is the use of thin interference films for *increasing* the reflectivity of a glass surface. In this case a requisite is a film which has a *higher* reflectivity than that of glass and can be vacuum deposited in a manner similar to cryolite. Such a material has been found in zinc sulphide for which the refractive index is $2\cdot37$.

Consider now the deposition of a $\lambda/4$ film of zinc sulphide ($\mu_2 = 2\cdot37$) on glass ($\mu_3 = 1\cdot5$). From the first face comes back about 16% because of the high refractive index. The residue passes on to meet the film-glass interface, and because reflection takes place from within a high-index to a low-index region an additional phase change of π now intervenes and the result is that the two reflected beams now *reinforce* each other. When account is taken of multiple reflections and when the two amplitudes are added it is found that the combined result is to produce a reflected beam 31% of the incident intensity.

Now the bigger the difference between μ_2 and μ_3 the more the reflectivity, and it is clear that use can be made of the low refractive index of cryolite by depositing a triple layer H_1-L_1-H_2 in which H is ZnS and L is MgF_2. For the first pair of films H_1-L_1 will give the above high reflectivity, then again the next pair H_2-glass will reflect part of the residue.

The whole can be repeated again and although exact calculation including multiple reflections is complex, it can be established that the reflectivities increase with the number of multilayers in accordance with the following table:

Layers	1	3	5	7	9
$R\%$	31	67	87	94	97

However, not only are the reflectivities high, *but the absorptions are very low*, usually less than 1%, and this combination of high reflectivity with low absorption has, as will be shown later, valuable application.

As in the case of the anti-reflection films, the multilayers act specifically for one wavelength and the enhanced reflectivity falls off fairly rapidly with alteration in wavelength. However, in both reducing and augmenting reflectivity one is at liberty to select any value of λ for the $\lambda/4$ films, hence one can, if so required, produce reduced or increased reflectivity for any required specified wavelength. The mechanism does not operate over a wide wavelength range for any given selected multilayer.

Liquid interferometers

It was pointed out by Rayleigh in 1893 that a water surface could be used as a natural flat for interferometry. He immersed a flat glass plate *below* the liquid, a short distance below the surface. Interference fringes were produced between the surfaces of the water and glass plate. Rayleigh was able to demonstrate the surface-tension curvature effect on the liquid surface produced near the boundaries of the container.

The flatness of a liquid surface will be affected by: (a) curvature of the earth, (b) surface tension effects at boundaries, (c) vibrations. The curvature of the earth causes a projection of only $\lambda/500$ (for yellow light) over a 25-cm diameter disc and can thus be disregarded. The surface-tension effect is found by experiment to vanish at about 2·5 cm from the boundary, so that, provided the diameter of the dish is 5 cm greater than that of the surface to be matched, the liquid is effectively perfectly flat. Vibrations are reduced if a viscous liquid, for example, medicinal paraffin oil, is used.

Liquid interferometers can be used for two distinct purposes, namely either (a) for examining slight displacements of liquid surfaces or (b) for using liquid surfaces as reference "flats" to test large workpieces. A typical example of the first case was described by Fabry (1927) who uses the following method for making measurements on the diamagnetism of liquids. A long-focal-length lens rests on the base of a glass dish, and below this is a magnet. Liquid is poured into the dish until it just covers the lens. Newton's rings can be seen, formed in the liquid film. If this film changes in thickness then the passage of each fringe corresponds to a thickness change of $\lambda/2\mu$ where μ refers to the liquid.

When a magnetic field is applied and a diamagnetic liquid is used, the level drops near the lens and rises some distance elsewhere. The outer rise is small if the dish is large, but can in any case be determined by an auxiliary interference system (if such accuracy be required). With water, a field of 250 gauss can produce a single fringe shift and 2000 gauss produces a 60-fringe shift, so sensitive is the arrangement that if the ambient air is replaced by oxygen a noticeable increased effect is found.

The second method of using a liquid surface interferometer, that in which the surface is used as a test plane, was described by Barrell and Mariner (1940). This is essentially a reflection Fizeau fringe system, arranged as in Fig. 6.17, but used for the examination of a metal plate.

The polished plate to be tested is covered by a 3-mm thick level of liquid paraffin. With tilting screws it is adjusted parallel to the liquid surface and Fizeau fringes formed between the liquid surface and the plate revealing any deviations from flatness. The thin liquid film helps to damp vibrations.

For the examination of glass plates it is not possible to use an *immersion* method, for the refractive indices of glass and liquid are so near as to make the intensity reflected from the interface to be too small. In this case the plate to be tested is held *above* and *outside* of the liquid surface, whilst the base plate of the container of the liquid is now tilted to throw away the reflected beam from this plane, to miss the eye.

The particular value of this method of testing flats is that large areas can be examined easily without auxiliary flats. Normally, three separate "flat" surfaces are required to confirm flatness, for it is clear that two surfaces, one slightly convex, the other slightly concave, will fit together parallel to be everywhere equidistant and thus both appear plane. If plate 1 is concave and plate 2 convex, then the above condition holds. If plate 3 is also convex, then 1 and 3 will again match, but 2 and 3 will not match. Hence to establish planeness it is necessary that three separate flats should match in pairs. The cost of preparing large proof planes, as they are called, is considerable, and a liquid interferometer is an inexpensive approach, provided the standard required is not too severe. There is no difficulty in detecting errors of $\lambda/10$ over a disc of 10 cm diameter.

CHAPTER 7

FRINGES OF EQUAL INCLINATION

We have been concerned up to now with the treatment of interference effects which occur between surfaces inclined at a small
angle to each other.

In the governing condition $n\lambda = 2\mu t \cos \theta$ we have generally
restricted the incidence to normal ($\cos \theta = 1$) and successive
fringes, appearing to contour regions of constant μt, have been
called fringes of equal thickness. Now we shall consider strictly
parallel surfaces, i.e. μt constant, and discuss the fringe patterns
which arise, when θ varies. These, it will shortly be seen, can be
called *fringes of equal inclination*. For the production of fringes
of equal thickness a parallel beam (θ constant) was secured by
using a point source at the focus of a lens. On the contrary, for
viewing fringes with strictly parallel surfaces, i.e. fringes of equal
inclination, a range of θ values is needed, and to do this, instead
of a point source, an *extended* source is required. This may be
either a simple broad diffuse source, but preferably should be in
the focal plane of a lens.

To simplify diagrams we shall consider transmission fringes,
disregarding the effects of intensities on amplitudes. First, the
shape of the fringes and then their location will be discussed. In
the formula $n\lambda = 2\mu t \cos \theta$ it is seen that successive fringes will
appear for those values of θ for which n is an integer, the particular
values being fixed by μt and λ. For particular values of θ we can
visualize the position as in Fig. 7.1. For a three-dimensional
picture it is necessary to rotate the diagram about AB. All the
light incident at angle θ_1 falls along the surface of a cone of semi-
angle θ_1. Since all this contributes to one fringe, clearly the
fringe is a circular ring. For some other angle of incidence θ_2
another ring appears. It is clear why an extended source is
needed if the source is in a lens focal plane. A point source
would give one angle of incidence and thus a complete set of rings
would not form.

Since there may be some difficulty in recognizing why circular

77

rings are formed by parallel plates another simple approach may be considered. To an observer it would appear that the point A and its image A′, produced by reflection, together produce the interference fringes. In other words, one is viewing interference from two sources in line of sight. It has already been demonstrated in earlier chapters that this leads to formation of *ring* fringes.

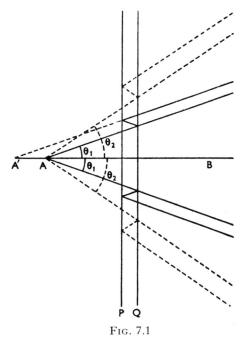

FIG. 7.1

It is clear from Fig. 7.1 that the pairs of beams which combine to interfere emerge parallel, because the two surfaces PQ are parallel. These beams thus can only be brought to interfere by a lens, for at the focus the parallel beams are united. Alternatively, if the eye is relaxed on infinity the beams meet on the retina. Thus the rings are described as being *localized at infinity*. This being so, a ring "diameter" is characterized by its *angular* diameter 2θ. If the rings are thrown on to a screen placed at the focal plane of a lens of focal length f, then the *linear* diameter on the screen is $2\theta f$. Such rings given by parallel plates were first found in 1849 by Haidinger, using a plate of mica, and are often termed Haidinger fringes. In reflection, since two equal beams were involved, good visibility results, but with low-reflecting surfaces like glass or mica, the rings can only barely be detected in transmission.

Properties of the Rings

Whether using an air film between glass plates, or with a single sheet of material like mica, in both cases there is a phase difference introduced between the beams as well as the metrical path difference. Hence in reflection for *dark* fringes we have $2\mu t \cos \theta = n\lambda$ and for bright fringes $2\mu t \cos \theta = (n+\frac{1}{2})\lambda$. In transmission the complementary condition exists.

Consider the dark rings in reflection, for an air film ($\mu = 1$) of thickness t. Then, writing n_0 for the order of interference at the centre (normal incidence, $\cos \theta = 1$) gives $n_0 = 2t/\lambda$. This may well be a large number. For example, for $\lambda 5000$ and $t = 1$ mm, n_0 is 4000. For any ring n (defined by θ), then $n = n_0 \cos \theta$. Writing ϕ as the angular diameter of the ring for which θ is the angular radius,

$$n = n_0 \cos \phi/2 = n_0(1 - 2 \sin^2 \phi/4)$$

In all cases of t, except t very small, the value of ϕ is such that ϕ can replace $\sin \phi$, hence $n = n_0(1 - \phi^2/8)$. This gives $\phi = \sqrt{\{(8/n_0)(n_0 - n)\}}$ as the angular diameter. Thus for a lens focal length f the linear diameter is $f\phi = \sqrt{(8f^2/n_0)(n_0 - n)}$. This can be written $2f\sqrt{(\lambda/t)}.(n_0 - n)$.

For simplicity consider a value of t such that n_0 is a whole number, then successive rings occur for values of n diminishing by 1 such that $n_0 - n$ is successively 1, 2, 3, 4 .... Thus ring diameters, exactly as in Newton's rings, are proportional to the roots of the natural numbers. The diameters are also proportional to $\sqrt{\lambda}$ and inversely proportional to $\sqrt{t}$.

To take a specific example with a lens of focal length 25 cm, $\lambda = 5000$ Å and $t = 1$ mm, the diameter of the first ring on the screen is about 1·1 cm. One can, under suitable conditions, obtain Haidinger rings over much longer path differences, even over 100 mm, for which thickness the diameters of the ring patterns are only one-tenth of the above.

Going back to the 1-mm gap, the angular diameter of the first ring is 0·044 radian, i.e. 2·5°, that of the second ring being 3·5°.

For the longer gaps the angles are proportionately less, from which it is clear that the approximation of ϕ for $\sin \phi$ is quite valid.

It will be seen that the production of fringes does not require small gaps, the small path differences, even with long gaps are a consequence of the very slow rate of diminution with angle of $\cos \theta$ in the neighbourhood of $\theta = 0°$.

There is a most important difference from Newton's rings, in that in Haidinger's rings the order of interference is a *maximum* at the centre (n_0) and *diminishes* progressively for the outer rings. In Newton's rings the converse is true.

In general, the order of interference at the centre n_0 is not a whole number, the first whole number, n, applying to the first *complete* ring. The centre order n_0 can be written $n+\epsilon$ where ϵ is a fraction. The general expression for the diameter of the first ring now becomes:

$$D = 2f\sqrt{\epsilon} \cdot \sqrt{(\lambda/t)}$$

For the $(p+1)$th ring we have:

$$D_{p+1}^2 = (4f^2\lambda/t)(\epsilon+p)$$

To evaluate ϵ it is only necessary to measure adjacent ring diameters D_{p+1} and D_p, whence

$$\frac{D_{p+1}^2}{D_{p+1}^2-D_p^2} = \frac{\epsilon+p}{\epsilon+p-(\epsilon+p-1)} = \epsilon+p$$

Brewster's Fringes

Discovered by Brewster in 1817, fringes of equal inclination can be obtained by means of two inclined-plane parallel plates of effectively the same thickness. They remained for long somewhat ignored, but adaptions of the system have played increasingly important parts in interferometry.

The system (like Newton's and Haidinger's rings) is also best seen in reflection, but again for diagrammatic simplicity the transmitted system will be considered. (It will indeed be shown later that great improvements in definition obtain with all these systems by covering the surfaces involved with high reflecting silver films. In these cases one does indeed frequently employ transmitted light, so that the treatment of the transmission fringes turns out in modern practice to be by far the more important.)

Consider two similar parallel plates A and B as in Figs. 7.2 (a), (b), (c). Suppose two parallel superposed light rays P and Q strike the plates at near normal incidence. (They are displaced for clarity.) We see that there exists a whole set of arrangements in which each ray goes through the same path length as the other, and, of these, three are illustrated. Indeed, multiple reflection can lead to any number of such pairs, in all of which the path length is the same for each member.

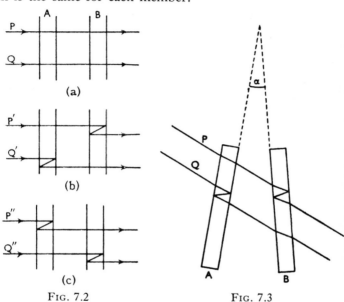

Fig. 7.2   Fig. 7.3

Now as in Fig. 7.3 let the plates be inclined to each other to include an angle α, both plates being set in vertical planes and let light be incident at an angle other than normal. Let r_A, r_B be the angles of refraction in the two plates depending on the angle of incidence and on α. As already established the path difference between direct and once-reflected rays in the first plate A is $2\mu t \cos r_A$, with Q lagging behind P. A further path difference $2\mu t \cos r_B$ is introduced by plate B, but now with P lagging behind Q. Hence the whole path difference is

$$D = 2\mu t(\cos r_A - \cos r_B)$$
$$= 4\mu t\left(\sin \frac{r_A + r_B}{2}\right)\left(\sin \frac{r_B - r_A}{2}\right)$$

A simpler condition arises when symmetry is arranged such that the incident light is incident to bisect the two normals, i.e. incidence is $\alpha/2$ on both plates and $r_A = r_B$.

Under this symmetrical condition we see that D is zero, no matter how thick the plates.

There is reinforcement at this point, and this is true for all wavelengths so that with white light a reinforcement also occurs. Now suppose we have a range of angles of incidence on either side of $\alpha/2$, then the path difference is $2\mu t(\cos r_A - \cos r_B)$ and this increases fairly uniformly with increasing angle. At positions where this equals $n\lambda$ fringes form. Since the two beams which contribute to a fringe emerge parallel, the fringes are at infinity. They are straight-line fringes of equal inclination, parallel to the edge of the wedge formed by the two inclined plates.

With white light one sees a central bright fringe and a few coloured fringes on either side, with monochromatic light quite an extensive fringe system becomes visible. It will be seen later that Brewster fringes, modified by silvering the four glass surfaces, have important applications. We shall consider here a modification, due to Jamin in 1856, which was one of the earliest important interferometers and is an instrument still in regular use.

The Jamin Interferometer

The Jamin interferometer, until the advent of the Rayleigh interferometer, was the most important instrument for the measurement of the refractive indices of gases. Essentially it makes use

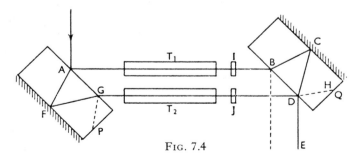

FIG. 7.4

of Brewster's fringes with the special condition of 45° incidence. The arrangement is shown in Fig. 7.4. The essential components are two identical plane-parallel thick plates P and Q (cut from the same block) at least $2\frac{1}{2}$ cm thick. Let the plates be set parallel.

The hatched faces of the blocks are heavily silvered. A *narrow* parallel light beam is sent from the source on to P. Consider one ray. It meets the unsilvered face at A, there is division of amplitude. A somewhat weak beam travels to B, there a small fraction is reflected (dotted) but this will be disregarded, for the main part passes on to C, is practically all reflected and emerges at D and goes to E. The weak fraction reflected back (dotted) at D is to be disregarded too, for it will not be used. The main part of the incident ray at A refracts to F, passes to G, thence to D at which part is reflected to E and part refracted in the direction H. The tubes T_1, T_2 have apertures which act as stops controlling the width of the incident pencil such that neither of the dotted beams at B or H can enter the viewing telescope. Thus we have essentially a Brewster system in which the two paths ABCD and AFGD are equal and the two beams arriving at E have the same intensity. Since path difference is zero, white-light fringes will appear if an extended source is used. By suitably tilting one plate, the fringes, which are at infinity, can be made horizontal in a viewing telescope.

The plates are thick in order that the two beams AB and GD should be well separated. The beams pass respectively down the two tubes which are initially evacuated and then the gas to be studied is introduced into one of them. Fringe displacements, as in the Rayleigh interferometer, are measured in terms of one wavelength by means of the Jamin compensator, IJ, already described in connection with Rayleigh's instrument. The calculation for refractive index is carried out in a similar way.

The Rayleigh instrument is the more sensitive for several reasons. First and most important is the fact that the fiduciary fringe-matching system in the Rayleigh refractometer is superior to that in setting a cross wire. Secondly, there is less likelihood of temperature drift causing a serious effect with Rayleigh's instrument, since both the gas fringes and the fiduciary system will tend to drift simultaneously. Thirdly, the fringe localization with the Jamin system is affected if the plates are not strictly parallel. Fourthly, one has quite a considerable mass of glass to maintain at a steady temperature and there is always the possibility that the block nearest the light source might warm up.

A closely allied system, the interferometer of Mach and Zehnder, will be described in a later section.

CHAPTER 8

MICHELSON'S INTERFEROMETER

IT is probably correct to say that no single instrument has more profoundly affected modern physics than has the Michelson interferometer, which is an instrument employing division of amplitude. For with it Michelson laid the experimental foundation to the theory of relativity, established the existence of hyperfine structure in line spectra (now of much importance in nuclear physics), evaluated the metre in terms of light waves, thus laying the foundation of modern precision metrology and measured the tidal effect of the moon on the earth. With slight modifications the instrument has found widespread use in optical workshops for lens and prism testing, it has been turned into an

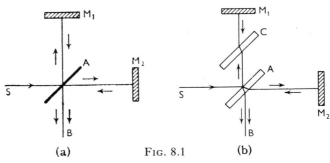

(a)   FIG. 8.1   (b)

interference microscope for studying the microtopography of surfaces and has been used as a refractometer. It has indeed found employ in innumerable investigations covering very many different fields of optics and physics.

So important an instrument deserves detailed consideration. It owes its success largely to its versatility, simplicity and stability. We shall find that numerous variants have been at times developed but shall restrict ourselves to the simpler types, discussing the principles and reviewing in detail a few only of the many applications that have been made.

In its first form the instrument was described by Michelson

in 1881. It consists, in essential, of a half-silvered mirror, which is a very thin deposit of silver, transmitting and reflecting equal amounts, which divides a beam into two parts of equal intensity, one transmitted, the other reflected. These two beams meet mirrors at practically normal incidence, return to the beam splitter and recombine. The basic arrangement is that of Fig. 8.1 (a).

Light from a source S is incident at 45° on the front half-silvered mirror A. The beam amplitude divides and half goes to M_1, the other half to M_2. The beams return to A and from thence to the eye at B, where fringes are seen. Either M_1 or M_2 can move on a carefully machined slide. Both mirrors are provided with three tilting screws.

In practice the reflector A was formerly a thin silver film on glass, nowadays it is usually a dielectric film 50% reflector on glass. It will be seen later that it is required often that the optical paths AM_1A and AM_2A should be identical. From Fig. 8.1 (b) it is seen that the paths of the two beams through glass are not the same, hence a compensating plane parallel plate C of same thickness as A is introduced into the shorter path and if this is parallel to A, the paths are identical in the two arms. If the plate A is silvered on the back face, the compensator is placed in the arm AM_2.

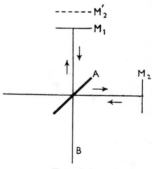

Fig. 8.2

Reverting back to Fig. 8.1 (a), from the viewpoint of an observer, the image of M_2 appears as in Fig. 8.2 to come from M_2'. The instrument is designed so that M_2 can move along accurately worked parallel metal slides, and with a good-quality screw the distance M_1M_2' can be varied at will. Both M_1 and M_2 have three adjusting screws for altering the angle of inclination between the two planes M_1 and M_2'.

The Nature of the Fringes

A detailed calculation of the nature of the fringes formed is some-what complex, but it is possible to arrive at their form by using the elementary considerations employed in Chapter 1. Opticall the system as viewed from B consists of two sources M_1 and M_2' behind each other. If the two image planes M_1 and M_2' are parallel, this is equivalent to sources in line behind each other and one sees the circular fringes characteristic of parallel plates. If M_1 and M_2' intersect, the cross-over is the position of zero path difference, and as this region is a straight line of intersection, white-light fringes can appear and will be straight lines, parallel to the line of intersection. They will be localized strictly at this intersection. If the two planes are inclined but do not intersect, i.e. encompass a wedge, then it can be demonstrated that one obtains curved fringes, sections of hyperbolae, the direction of curvature changing with the direction of the wedge apex. It is important to note that the fringes appear to be localized at the front mirror M_1.

We shall now consider the fringe pattern a little more analytically.

Theory of Fringe Shape

Let the two plane images M_1 and M_2' producing the interference be inclined at a small angle α as in Fig. 8.3. Let the eye be at O, and let R be the foot of the perpendicular from O on to the plane M_1. With OR $= z$, let the coordinates in the plane of the point P be x, y. Consider the effect on O of rays from P.

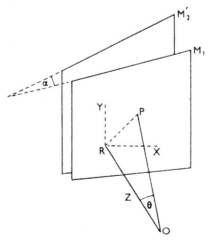

FIG. 8.3

To O, the angle of interference is θ and the path difference of rays from the front and rear planes reaching O from P is $2d \cos \theta$, if d is the separation of the planes at P.

Since $$OP^2 = (x^2+y^2+z^2)$$

then $$\cos \theta = z/\sqrt{(x^2+y^2+z^2)}$$

If the film thickness at R is e, that at P is $(e+x \tan \alpha)$. For a fringe, then,

$$\frac{2(e+x \tan \alpha).z}{\sqrt{(x^2+y^2+z^2)}} = n\lambda \qquad . \quad . \quad . \quad (8.1)$$

This expression governs the shape of the fringes.

Consider the special case in which the two planes intersect at R, then $e = 0$ so that

$$n\lambda \doteq \frac{2xz \tan \alpha}{\sqrt{(x^2+y^2+z^2)}}$$

If, further, z is large compared with x and y, this approximates to $n\lambda = 2x \tan \alpha = 2x\alpha$, for α is small. This is the equation of a straight line, so that for any given value of n, x is constant, which means that successive n values, i.e. successive fringes, give straight lines parallel to the axis. The fringes are equidistant, separated by the distance $\lambda/2\alpha$. They appear localized on the plane M_1.

This condition only holds when $z \gg x+y$. If the eye approaches the plane M_1, then $x+y$ is no longer negligible and, instead of a straight line, the equation for n becomes that of a hyperbola, in which the line of intersection of the mirror is the axis of the hyperbola. Thus according to the angle of tilt the fringes may be curved.

Consider now the special case in which the two planes M_1 and M_2' are parallel, i.e. $\alpha = 0$, then equation (8.1) reduces to

$$n\lambda = \frac{2ez}{\sqrt{(x^2+y^2+z^2)}}$$

Square this, giving:

$$(x^2+y^2+z^2) = 4e^2z^2/n^2\lambda^2$$

i.e. $$x^2+y^2 = \left(\frac{4e^2-n^2\lambda^2}{n^2\lambda^2}\right)z^2$$

If the eye is at a given place, z is constant, then the locus for n constant will be a fringe. As the surfaces are parallel e is constant, thus x^2+y^2 is constant, which is the equation of a circle. Fringes therefore, are circles, with centre at R and radii $(z/n\lambda)\sqrt{(4e^2-n^2\lambda^2)}$. For e less than $\lambda/2$ no rings form and one sees a region of regular illumination. When e is small the diameter of the first ring exceeds the aperture of the instrument and the field of view is effectively uniformly illuminated. Since the emerging beams are parallel the circles appear to be at infinity. Summarizing then, if the two planes are inclined and intersect, straight-line fringes are seen. If the plates are inclined but do not cross, then curved fringes result, sections of hyperbolae. If the plates are parallel circular fringes appear at infinity. All this is in accordance with earlier described elementary treatment of two sources in line.

In the above discussion only the optical and not the metrical separation of the two mirrors from the beam has been considered. It may be asked why the compensating plate has been introduced. Why not move mirror M_1 back by an amount equal to the extra optical path introduced by the thickness of glass supporting the beam-splitting film ? Indeed this can be done if the instrument is to be used only in *monochromatic* light, but such an arrangement fails with *white light* because of the dispersion of the glass. If the mirror is moved to give equal paths for say green light, the paths will no longer be equal for say red or blue light. The only way to secure equality in path for *all* wavelengths is the introduction of a compensating plate identical in thickness and material with the plate bearing the beam-splitter. This compensating plate can also be additionally effectively employed in some experiments for calibration purposes.

Michelson carried out a series of fundamentally important experiments with his instrument. First he studied the narrowness of spectral lines. Having found a sharp line he used this to evaluate the metre in terms of light waves. He made observations on the relative velocity of light, his results leading Einstein to formulate the Special Theory of Relativity. He carried out costly formidable experiments on the rotation of the earth and also on the tidal effect of the moon. We shall consider some of his experiments although not in the strict historical sequence.

The Structures of Spectral Lines

It has already been pointed out that Fizeau noticed the fluctuation in fringe visibility with sodium light when the components of a

Newton's rings combination were progressively separated. A corresponding effect appears with the Michelson interferometer, for as the movable plate is slowly retracted the fringe visibility changes. Michelson (1891) systematically developed this and was able to infer the existence in some cases of very close spectrum-line structures. Indeed, the very widths of spectrum lines themselves could be derived with what is effectively very high spectroscopic resolving power merely by studying fringe visibility.

The observations were conducted on circular ring fringes and the quantity to be determined, the visibility V is defined as

$$V = \frac{I_{max} - I_{min}}{I_{max} + I_{min}}$$

This was done by inspection, by comparing the appearance of the rings with patterns of known visibility. These patterns of known visibility were secured by placing a concave lens of crystalline quartz between crossed Nicol prisms. Such an arrangement produces circular interference rings due to the double refraction of the quartz. The visibility of such a ring pattern is determined by the angle, through which the quartz is rotated relative to the axes of the Nicols.

The visibility curve as a function of separation for a doublet is shown in Fig. 8.4 (a) and this is similar with that which can be calculated from the formula for fringe intensity $I = a_1^2 + a_2^2 + 2a_1a_2 \cos \theta$ which has already been discussed. It follows conversely that such an observed visibility on other lines implies the existence of a doublet in each case.

The problem of converting back from the visibility curve to determining the character of the original source is one of Fourier analysis. For a strictly monochromatic source the visibility curve would be a horizontal straight line. Any particular pattern or arrangement of separate wavelengths in a source will produce its own particular net effect on the visibility. Michelson invented a mechanical "harmonic analyser" which was a machine for giving an approximate solution to the conversion problem. A template is cut out which has the shape of the observed visibility curve. Rods of uniform length are arranged to rest on this template and, by an ingenious mechanical arrangement, the displacements of these rods above the horizontal are converted to give as a resultant the original line distributions in the form of curves.

For example, the visibility curve given by the red hydrogen line and shown in Fig. 8.4 (b) resulted in the harmonic analysis shown alongside. The source is found to be a doublet with components of separation 0·14 Å. Later instruments which do resolve and separate the components of the H_α line show that there are at least three components, but that the separation of the two principal lines is 0·136 Å, which closely confirms Michelson's deductions.

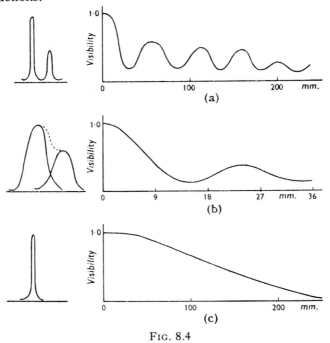

Fig. 8.4

Thus for the first time the detection of narrow spectroscopically unresolved structures in lines was achieved. There are two drawbacks to the technique. First, it is not possible easily to establish the phases as well as the visibilities. To do so one requires an approximately homogeneous source of the same wavelength. As a result of this the separation of components cannot be exactly computed, nor can it readily be established whether a component is on the long- or short-wavelength side.

An important result established by Michelson was the confirmation of the validity of the theoretical expression for Doppler width, proving that the line width is proportional to $\sqrt{(T/m)}$ in which T is the absolute temperature and m the mass of the emitter, in accordance with kinetic theory.

Of striking interest was the visibility curve and its analysis found for the red cadmium line 6438 Å, shown in Fig. 8.4 (c). The line in fact proved to be quite single (and later work has amply confirmed this). From the visibility curve the line shape could be derived, and this was found to follow closely the Maxwell distribution expected from the Doppler effect. The half-width, which is the width at an intensity half of the maximum, was 0·013 Å.

This line was selected by Michelson for the evaluation of the metre because good fringes could easily be secured over the 100-mm path *difference* which was to be used in that determination.

The line-width visibility curve of any line indicates the limit of path length over which fringes can be detected.

The Measurement of the Metre in Light Waves

By international agreement lengths in science were measured in terms of the standard metre. This is a bar of platino-iridium of X-section maintained at Sèvres, near Paris. On it, on polished sections, are marked two fine scratches made by diamonds, each about 5 μ wide or less. The metre was defined as the distance between these at 0°C. It has been established that a visual setting by a microscope cross wire can be made on the scratches with a standard deviation of 0·3 μ, which quantity is half a light wave of wavelength 6000 Å. Experiment shows that repeated observations to eliminate systematic error can lead to a standard deviation of 0·06 μ. It should be recalled that the resolving power of a microscope is about half a light wave.

There are two drawbacks to such a standard. First, there is the somewhat remote possibility of the standard being destroyed. (Modern warfare removes this conjecture from the category of highly improbable to that of the possible.) Second, there is always the very real possibility that the metal standard may change in length over long time periods due to slow crystallization effects. Indeed in the case of the Imperial Standard Yard, there is reliable evidence that this length standard decreased by 0·0002 in. during the first few years after its establishment in 1856.

Babinet, in 1829, was the first to suggest that light waves should be used as a natural and invariable unit of length. This has in fact not yet been adopted, but, as a near approach, Michelson first decided, and international agreement later concurred, that the existing standard metre should be evaluated in terms of light waves. If then the standard were destroyed, it could be accurately replaced. Further, if the length changed with time, redeterminations in terms of light waves should reveal this. In fact, during the past sixty years there have been no less than nine precision evaluations of the metre, and it is a tribute to the experimental brilliance of Michelson that the mean of these agrees with Michelson's early value to within 1 part in 16 millions.

To define the metre in terms of light waves requires the selection of a suitable wavelength and either the determination of the refractive index of air, under specified conditions of composition, temperature and pressure, or else a determination in a vacuum. In the first instance Michelson had suggested using the yellow sodium lines, but, after systematic examination with his interferometer, he selected the red line from the cadmium spectrum as a line of extreme sharpness, well suited to his purpose. It speaks volumes for the skill and foresight of Michelson, when it is recognized that this source has remained, up to within very recent years, unrivalled, and has only been improved upon as an incidental consequence of the recent development of the atomic energy piles, in a manner to be described later.

The wavelength of the red cadmium line in terms of the standard metre is the standard wavelength to which all other wavelengths are referred. Thus it is that the very high attainable precision of spectroscopy (about 2 parts in 10^8) is dependent basically upon the interferometric evaluation of the metre. It will be shown later also that the intercomparison of different wavelengths is secured too by interferometric means.

Michelson's Evaluation of the Metre

The source of light used by Michelson was a glow discharge through cadmium vapour in a Geissler tube maintained at 300°C. Improved sources are now available. A basic underlying difficulty was the fact that despite the relative sharpness of the source it is not possible to secure interference with it over a path difference of 1 metre. Interference over a path difference of only 20 cm is very difficult to establish and this is about the limit. Again, the number of red light waves in a metre turns out to be about $1\frac{1}{2}$ millions, so that a count of the fringes in a metre would involve the enormous task of counting 3 million fringes. Not only would this take up so much time as to make it virtually impossible to keep the physical conditions constant (counting one fringe per second would take 800 hours), but in addition it would not be humanly possible to make such a count without making many mistakes.

It was decided in the first instance to measure the waves in a 100-mm length by first subdividing this to 50 mm, then to 25 mm and so on down to eight subdivisions, the last of which $(100 \div 2^8) = 0.3906$ mm. Such a length contains some 600 light waves, i.e. a fringe count of some 1200 is needed, a relatively small number. The substandards, made as accurately as possible (Fig. 8.5), consisted essentially of mirrors M_1, M_2 which can be brought parallel by grinding and polishing contact pins. This is done in the

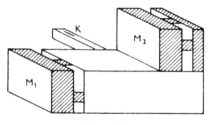

FIG. 8.5

interferometer. The experiment consists of the following sequence of operations:

(1) a count of number of wavelengths on the first (smallest) substandard;

(2) the determination of the difference between twice the first substandard and the next;

(3) repetition of this until the 100-mm substandard has been evaluated;

(4) the stepping up of the 100-mm standard ten times against itself, finally matching this against the scratch marks on the standard metre.

To reduce vibration and assist temperature control, the two arms of the interferometer were brought parallel by means of an extra mirror P as in Fig. 8.6.

The reference mirror L had a rectangular grid scribed on it, permitting fringes to be brought to any specified position in the field of view.

For any substandard, such as the first, the parallelism of M_1, M_2 was achieved thus. M_1L were set parallel to give ring fringes, and M_2 then adjusted for the same. All standards were set in a similar way and clearly M_1 and M_2 are both now parallel to L. The first and second standards (M_1, M_2) and (N_1, N_2) were placed as in Fig. 8.6, and the sequence of observations and settings was then as follows. The mirror L matches in four quadrants respectively with M_1, M_2, N_1, N_2.

(i) M_1 and N_1 were first brought into coincidence against L using white light, they only covered the lower half of L.

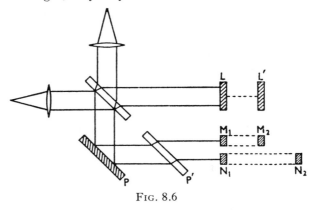

FIG. 8.6

(ii) N_1 was set parallel to L, but M_1 was slightly inclined to L to give vertical white-light fringes, placed with the central achromatic fringe to coincide with a selected scribed mark on L.

(iii) The reference mirror L was then slowly moved and the passage of circular red cadmium rings in N_1 counted. L was moved to L' at which point the left upper quadrant of L was now in coincidence with M_2. This was established by white-light fringes being brought to the same scribed mark in the upper left quadrant (for M_2) as in the lower left quadrant (for M_1). The separation M_1M_2 was not found to be an exact number of fringes, and the fraction was determined by noting the fact that the circular rings were not at the same phase at the beginning and end of the count. They were brought to the same state (i.e. same fraction at the centre) by introducing a fraction of an order through rotation of the compensating plate P' through an angle which was noted.

This was first previously calibrated, the rotation to produce a single fringe shift having been determined, and thus the fraction was evaluated.

In the experiment it was found that the first substandard showed 1212·37 fringes for the red line.

(iv) The next step was the comparison of twice the length of M_1, M_2 with N_1, N_2, achieved quickly as follows. With white light L, M_1, N_1 were brought into coincidence, and L tilted to give vertical white fringes on both M_1 and N_1. L was now moved to L′ to give white-light fringes on M_2. The whole unit M_1, M_2 was moved bodily until M_1 was in coincidence with L′, as shown by white-light fringes. If N_1, N_2 was exactly twice M_1, M_2 then M_2 and N_2 would coincide. There was a small difference which was to be determined by a count.

(v) Next L was moved until it coincided with the new position of M_2 (white light). Fringes were also seen in N_2 and the small difference between these determined as before with the compensating plate. It was thus only necessary to count for the first substandard and then to evaluate a small difference.

(vi) The sequence was repeated matching successive doubled substandards until the 100-mm standard was reached.

(vii) The final operation was the multiplication of the 100-mm substandard ten times and the matching against the metre itself. The final unit had on it (see Fig. 8.5) a projecting arm K on which was a fine scratch. Above the two scratches on the metre bar itself were fixed microscopes. K was brought into visual coincidence with one scratch. The mirrors M_1, M_2 were then stepped up ten times against L, using white-light fringes. The exact length of this ten-times multiplication was known, the error in the length of the 100-mm unit being now multiplied ten times in the final value. This distance was nearly, but not exactly, the metre. The unit was then moved until it coincided with the second scratch, the small difference between this position and the ten times position being measured by fringes. This gave the final value for the length of the metre bar.

The main error in the experiment was not in the interferometric determination but in the setting inaccuracy of the coincidences. This was increased by the fact that it was not permitted to handle the true standard but only a copy, which was matched by coincidence against the standard. The optical error was less than a fifth of a light wave.

The final result, given to 1 part in 2 millions, for air at 15°C and 760 mm was 1 553 163·5 waves of red cadmium light for the metre. The present accepted value will be discussed later.

Effects of Motion on Velocity of Light

Michelson's studies with his interferometer on the effect of motion on the velocity of light, culminating in the celebrated Michelson-Morley experiment (1886), have exerted a profound influence on modern physical thought, for it was on this basis that the whole edifice of the relativity theory was built. Michelson's experiments were a continuation of a series of curious

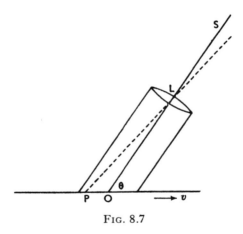

FIG. 8.7

results which had been accumulating from the beginning of the nineteenth century.

First there was the unexpected experiment due to Airy, on stellar aberration. In 1726 Bradley had discovered the aberrations in the apparent positions of stars produced by the movement of the earth in its orbit round the sun. The nature of this is shown in Fig. 8.7.

Suppose an observer O at rest, views a star in the direction θ with a telescope, the telescope will be pointed in the direction OS. Now, if the telescope is moving in the horizontal direction with velocity v, it will be necessary to point the telescope in the direction PL. For the observer will travel the distance PO in the time taken for the light to travel from L to O. Let t be this time and c the velocity of light. Then, closely enough, OL $= ct$ and OP $= vt$. From triangle OPL,

$$\sin \text{PLO}/vt = \sin \text{LOP}/ct$$

since v is small compared with c. This reduces to

$$\angle \text{PLO} = (v/c) \sin \theta$$

This angle is the angle of aberration, and v/c is called the aberration constant. The earth moves in its orbit with a velocity $v = 18\frac{1}{2}$ miles per second, and as c is 186 000 miles per second, then, very closely, $v/c = 0.0001$ radian, which is about $20''$ of arc. Over a whole year each star appears to describe a small circle because of the rotation of the earth round its orbit and measurement confirms the theory.

Now it was pointed out by Airy that the time taken for the light to travel down the telescope would be increased if the telescope were filled with a liquid with refractive index μ. The aberration constant v/c would then become $\mu v/c$, so that, for water, the aberration might be expected to increase in the ratio $4/3$. Airy carried out the experiment *with the surprising result that the aberration did not in fact alter.*

This experimental paradox was at once accounted for by an earlier theory of Fresnel (1818), who, on classical grounds, assumed that a moving medium carrying light vibrations (call it ether if you will) entrains or traps the ether which is thus carried along by the moving matter.

Fresnel had propounded an elastic theory of propagation of light, showing that the velocity c is given by $c = \sqrt{(E/\rho)}$ in which E is the elasticity and ρ the density of the ether. Fresnel assumed that ρ varied in different materials but not E. Although this mechanical view is completely outmoded, yet it does lead to an explanation of the difficulty in Airy's experiment and also explains later work. It is introduced here merely because of its simplicity and is, of course, now replaced by electromagnetic theory. It is a crude but useful model.

Fresnel argues as follows. Let a glass block of 1 sq. cm cross-section move through the ether with velocity v. Postulate that the ether density ρ_1 in the block exceeds that in the vacuum, ρ. Clearly the ether in the block must move with it to some extent, for if not it would be left behind and the ether density in the glass would fall to ρ.

Let v_1 be the velocity of drift of the ether. As the block advances into the ether the quantity ρv enters at the front face, but at the back face the quantity $\rho_1(v-v_1)$ leaves. These must be equal otherwise there would be total loss or infinite accumulation, hence

$$\rho v = \rho_1(v-v_1)$$

giving $\qquad\qquad v_1 = v(1-\rho/\rho_1)$

Now since the velocity of light is given by $\sqrt{(E/\rho)}$ and as by assumption E is constant it follows that $\rho_1/\rho = \mu^2$ (where μ is the refractive index) giving as the drift velocity of the ether the value:

$$v_1 = \left(\frac{\mu^2-1}{\mu^2}\right)v$$

The quantity $(\mu^2-1)/\mu^2$ is called Fresnel's dragging coefficient. For air, μ is sufficiently close to unity to make this zero.

Now it can be shown from simple geometrical considerations, without adopting any particular theories of wave propagation, that Airy's null result leads to Fresnel's drag formula, as a necessary consequence.

If the Fresnel drag coefficient be introduced into the calculation of the aberration, there emerges the fact that the *aberration is the same with or without water in the telescope*. Thus, conversely, Airy's negative result confirms the validity of the Fresnel coefficient.

So important is this, and so difficult was it to appreciate that Fizeau in 1859 carried out an ingenious interferometric experiment to test directly the validity of the drag formula.

Fizeau's Velocity-drift Experiment

Fizeau arranged to secure interference from two beams of light passing down columns of water moving in opposite directions, using the set-up in Fig. 8.8. From a slit source S a parallel beam passes through two slits A, B, the beams meet at a mirror M. On their way they pass through tubes through which water flows, one beam going with the other against the flow. (The system is effectively that later employed in the Rayleigh refractometer, and has been described earlier.)

Fringes could have been seen at M, but it is advantageous to use a mirror M and reflect back the light such that their paths interchange. This doubles the path length in each direction and also eliminates compression errors on refractive index, due to flow. Fringes are observed at E by the arrival there of the two beams.

According to the Fresnel hypothesis the beam entering A which initially opposes the current is retarded because of the ether drag by the water in both paths A to M and also in M to B, whilst the ray entering at B is similarly advanced in both its paths B to M and M to A.

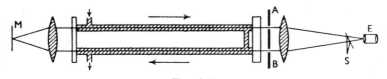

FIG. 8.8

Suppose v is the velocity of liquid flow and L the length of the tubes (which are sealed by plane parallel plates). The velocity of light in water when at rest is c/μ, and, if v is the water velocity, the Fresnel drag velocity is Kv (where $K = (\mu^2-1)/\mu^2$). The two velocities through the water in motion are then $(c/\mu)+Kv$ and $(c/\mu)-Kv$. If the tube length is L the time difference becomes:

$$\mathrm{d}T = \frac{2L}{(c/\mu)-Kv} - \frac{2L}{(c/\mu)+Kv}$$

Thus
$$\mathrm{d}T = \frac{4LKv}{(c^2/\mu^2)-K^2v^2}$$

The second order quantity K^2v^2 can be neglected compared with c^2/μ^2, hence

$$\mathrm{d}T = 4L\mu^2Kv/c^2$$

If T is the period of the light vibration, then $\mathrm{d}T/T$ is the number of fringes this time would correspond to. Calling this $\varDelta$, then $\varDelta = \mathrm{d}T/T$, which equals $(c/\lambda)\mathrm{d}T$. Substituting gives:

$$\varDelta = 4L\mu^2Kv/\lambda c$$
from which
$$K = \varDelta\lambda c/4L\mu^2v$$

The value of $\varDelta$ found by Fizeau, when substituted, gave $K = 0\cdot43$. But this is almost exactly the predicted Fresnel value for water, for, with $\mu = 4/3$,

$$K = \{(4/3)^2-1\}/(4/3)^2 = 0\cdot437$$

Although Fizeau's experiment gave a clear decision yet there were several inherent difficulties. Thus fine slits were needed, and this restricted light intensity. Then again the two slits A and B must be fairly close together, necessitating narrow-bore tubes, thus making it difficult to secure fast flows and also introducing flow variations over the cross-section of the light beam. Because of the importance of the experiment it was repeated by Michelson and Morley (1886) using a modification of the Michelson interferometer.

Michelson-Morley's Modification of the Fizeau Drift Experiment

The arrangement used is shown in Fig. 8.9. The beams are intense and are so widely separated that big-bore tubes can be used. A complex water circuit was designed so that current flow could be reversed and thus the effect doubled.

The velocities of flow in the tubes were directly measured with a Pitot flow tube. With a tube length of 6 metres and a flow velocity of 7 metres per second the observed displacement was 0·9 fringe.

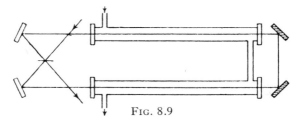

FIG. 8.9

This leads to $K = 0.434 \pm 0.02$, the predicted value given by the Fresnel formula being extremely close to this (0·437). A more detailed theoretical analysis by Lorentz takes into account the dispersion of the water and gives:

$$K = \frac{\mu^2 - 1}{\mu^2} - \frac{\lambda}{\mu} \frac{\mathrm{d}\mu}{\mathrm{d}\lambda}$$

If dispersion values are introduced, then for yellow light the value of K becomes now 0·451. This is still within the experimental value given by the drift experiment.

Thus, taken together, the aberration experiment of Airy and the drift experiments of Fizeau and of Michelson all confirm the validity of Fresnel's equation. Since $\mu = 1$ for space free from matter (or effectively so for air) it follows as a consequence that *there should be zero drift velocity*, i.e. *no relative motion between the earth and ether.* Thus, indeed, the velocity of light determined in air should not reveal any relative earth motion.

The Michelson-Morley Ether Drift Experiment

Maxwell pointed out that although no *first-order* effect (which is that sought for in the previous experiments) can be revealed, yet there were reasons for believing that *second-order* effects, of the size of the *square* of the aberration could be observed. If an experiment were sufficiently refined, such an effect should be observed. As the aberration constant is 1 in 10^4 then the square of the aberration is an effect of only 1 in 10^8. Yet there are about 10^8 light waves in a distance of 50 metres. If an interferometer can be set up with very long paths of this order (by using repeated reflections from mirrors), the expected effect should be detectable.

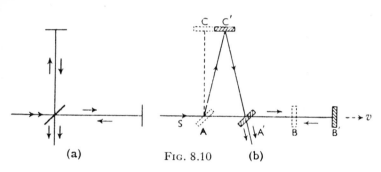

(a)   FIG. 8.10   (b)

Consider an interferometer set up as in Fig. 8.10 (a) when the system is at rest.

Let the two arms be of equal length l, and let the instrument now move with velocity v in the direction AB (Fig. 8.10 (b)). Because of the velocity displacement the mirror C has moved to C′ during the time taken for the light from A to reach it. Thus the light must be incident slightly off normal. For the horizontal arm the mirror B has moved to B′ by the time the light reaches it and A to A′ by the time the beam has returned.

Let T be the time for the beam to travel from A to B and T_1 that to travel back from B to A, in the position A′ when it is reached.   As the apparent velocity $A \to B$ is reduced from c to $c-v$, then

$$T = l/(c-v) \quad \text{also} \quad T_1 = l/(c+v)$$

The total time for the double journey is then

$$\frac{l}{(c+v)} + \frac{l}{(c-v)} = \frac{2lc}{(c^2-v^2)}$$

The distance travelled is c times this, i.e.

$$\frac{2lc^2}{c^2-v^2} = \frac{2l}{1-v^2/c^2}$$

or, as v/c is small, this is closely enough

$$L_1 = 2l(1+v^2/c^2)$$

This is to be compared with the length $2l$ which is traversed in a system at rest.

Now consider the light path in the direction AC'. As $AC = l$ and $CC'/AC' = v/c$, then, since

$$AC^2 = (AC')^2 - (C'C)^2$$
$$= (AC')^2(1-v^2/c^2)$$
$$\therefore \ AC' = l/(1-v^2/c^2)^{1/2}$$

i.e. to a first approximation

$$AC' = l(1+v^2/2c^2)$$

Since the total path is $AC'A'$, this equals

$$L_2 = 2l(1+v^2/2c^2)$$

Thus L_1 and L_2 are not the same, the difference being lv^2/c^2.

Now consider the effect of v as that due to the earth's motion, then v/c is the aberrational constant (10^{-4}) and the predicted effect is therefore indeed second order, for it is proportional to v^2/c^2. If l is 10 metres, the expected optical difference in the paths is 2×10^{-5} cm, which is two-fifths of a green light wave, an amount not difficult to observe with Michelson's interferometer.

The effect is double if the apparatus be turned through 90° so that the vertical and horizontal directions in the diagram are interchanged.

The interferometer system used was mounted on a stone block, $1\frac{1}{2}$ metres square and 25 cm thick, resting on an annular wooden ring and the whole was floated on mercury.   At each corner were placed mirrors and near the centre was placed the beam-splitter with compensating plate. The mirrors were arranged so that there were seven traverses there and back in each arm, a total path of about 11 metres resulting. If one considers motion of the earth only in its orbit, this leads to a predicted displacement of 0·4 fringe.

When the experiment was performed the stone was set into slow rotation and it was established that instead of the predicted 0·4-fringe displacement, *no displacement greater than* 0·01 *fringe was found.* This was a most unexpected result. There remained the possibility that the velocity v at the time of the experiment might have been accidentally small because of combined motions through space of the earth and the solar system itself. To obviate this, measurements were carried out at different times over the year but all confirmed a negative result.

The whole grand superstructure of relativity, with its tremendous implications ranging out, on the one hand, from theories of world cosmos, to on the other hand theories of the structure of the nucleus of an atom, is based upon the negative result in this experiment. Now it is notoriously difficult to prove a null result, and ever since the initial experiment there have been numerous repetitions, with increased refinements. Miller (1922) increased the path to 65 metres and again failed to find the predicted displacement. He did, however, report an apparent small drift, but his path was in air and calculation shows that a change of temperature of only 0·001°C or of pressure of 0·05 mm of mercury would account for the slight effect he found. Too long a path length is a disadvantage, hence Kennedy (1926) reduced the path length to 4 metres. Further, he replaced the air by helium, which has a lower refractive index than air so that pressure and temperature disturbances are less effective. Since a displacement of only some 0·15 fringe is expected, Kennedy aimed at increasing the sensitivity of the interferometer. This he achieved by building up (by deposition of silver) a step over half of one of the end mirrors of the interferometer. The height of this was only $\lambda/20$. The fringe pattern seen showed a discontinuity of one-tenth order on passing this step. If the fringes are broadened out to a sensitive uniform tint position, then one can establish a uniform bright field of view on one side of the dividing line, and a less bright field on the other. The fringes can also be displaced to reverse this situation. By judicious placing of the fringes one can secure uniform tint, both sides being off a true fringe zero, or maximum. A very slight displacement will then cause one half of the field to brighten and the other to darken. It is considered that a displacement of 0·001 of a fringe can be detected because the eye is very sensitive to the matching of equal tints.

With this greatly improved sensitivity Kennedy, and later Illingworth, and also several others, have confirmed the null result found originally by Michelson and Morley.

The Relativity Explanation of the Null Result

In 1905 Einstein showed that the observed null result follows as logical consequence of his relativity theory. Experimentally the Michelson-Morley experiment really establishes that the measured velocity of light is the same whether referred to the direction of motion, or at right angles to it. *This velocity as measured appears then to be independent of the velocity of the observer.* This is adopted as the fundamental basic statement of relativity. The

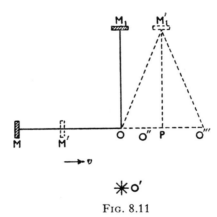

Fig. 8.11

rest is effectively a logical and algebraic deduction. Thus consider now the Michelson-Morley experiment in the light of this newly recognized property of the velocity of light. The diagram is slightly less confusing if the mirror is brought over from the right to the left.

Let M, M_1 be two mirrors equidistant from an observer O. Let M, M_1 and O be at rest with respect to each other and let O and O' have a relative velocity v in the direction MO. O sends out light signals simultaneously to M and M_1, and, as OM = OM_1 the signals return simultaneously.

Let the observer O' know that the observer O received the returned signals at the same instant. He is not aware of the fact that OM = OM_1, since the effective velocity of the light may be different in the two paths. Let him call the length in the direction of motion λ and that in the direction perpendicular to the motion l. Let the full lines in the figure represent positions relative to O' initially, and dotted lines subsequent positions as seen by this observer. We shall follow out the deductions which O' would make on the assumption that the velocity of light, with respect to himself, is independent of direction.

According to the observer O' the length MM' equals vt_1, and OM' equals ct_1 if t_1 is the time taken for the light signal to go from O to the mirror M, which by this time has moved to M'.

Now
$$OM' = OM - MM'$$
i.e.
$$ct_1 = \lambda - vt_1$$
giving
$$t_1 = \lambda/(c+v)$$

If t_2 is the time taken for the light to return from M' to the observer who is now at O''', then $M'O''' = ct_2$. Since $M'O'' = \lambda$ and as $O''O''' = vt_2$, then

$$ct_2 = \lambda + vt_2$$
i.e.
$$t_2 = \lambda/(c-v)$$

The total time taken for the light to go to the mirror M and come back is $T = t_1 + t_2$.

Substituting gives:

$$T = \{\lambda/(c+v)\} + \{\lambda/(c-v)\}$$
$$= 2c\lambda/(c^2 - v^2)$$

So far we have only considered the signal sent in the direction of motion, we shall therefore now examine the behaviour of the signal sent in the perpendicular direction. OM_1' is equal to $O'''M_1'$. As the total time taken for the double journey is T, *because the signals are received at the same time*, the time to travel over OM_1' is $T/2$. The distance $OM_1' = cT/2$, and as O has travelled to P during this time ($T/2$) it follows that OP = $vT/2$. Since $OM_1'P$ is a right-angled triangle,

$$OM_1'^2 = OP^2 + M_1'P^2$$
hence
$$c^2T^2/4 = (v^2T^2/4) + l^2$$
giving
$$T = 2l/(c^2 - v^2)^{1/2}$$

The value just derived for T can now be equated to that already found for the other arm, i.e.

$$2\lambda c/(c^2 - v^2) = 2l/(c^2 - v^2)^{1/2}$$
whence
$$\lambda = l(1 - v^2/c^2)^{1/2}$$

Thus, according to the observer O', the length l in the direction of motion contracts to λ where $\lambda = l(1 - v^2/c^2)^{1/2}$.

Since the length in the direction of motion appears to contract by an amount $lv^2/2c^2$ the total path there and back appears to have diminished by lv^2/c^2. *But this is exactly the increase in path predicted by the non-relativistic former calculation.* Thus the relativistic contraction exactly compensates the predicted non-relativistic increase, *hence a null observational result is to be expected.*

Amongst the many ramifications of this relativity theory a different (more refined) law emerges for the addition of velocities. When two velocities v_1, v_2 are added, instead of obtaining the classical formula $v = v_1 + v_2$, one readily obtains

$$v = \frac{v_1 + v_2}{1 + v_1 v_2/c^2}$$

Since most velocities (v_1, v_2) are small compared with c, for all physical measurement (other than those involving very high velocities), this reduces to $v = v_1 + v_2$. There are, however, two interesting properties of this law of addition of velocities. For let v_1 be the velocity of light, i.e. c, and let us add to this a velocity v_2, then the resultant velocity is

$$v = \frac{c + v_2}{1 + c v_2/c^2}$$

which reduces to c. The formula is thus entirely consistent with the basic assumption, for the addition of a velocity v_2 to the velocity of light c leads to the paradox that the result is still c. The velocity of light can thus be *approached* but *never exceeded*. Indeed, if a body could have the velocity of light, its length parameters, being reduced in the ratio $\sqrt{(1 - v^2/c^2)}$ would become zero. Also, since a further deduction shows that a rest mass, m, becomes $m/\sqrt{(1 - v^2/c^2)}$ at velocity, its mass at velocity c would become infinite. All these deductions show that the velocity of light cannot ever be attained by matter. Indeed, physically speaking, a prime difference between matter and radiation is that the former can never have the velocity of light, whilst the latter does have it.

Finally, taking the formula for addition of velocities, consider the Fresnel water-flow experiment, adding a flow velocity v_1 to that of the light, which in the medium has the value c/μ for its velocity v_2 because of the refractive index of the water. Then the resultant velocity of the light, according to the addition formula, is:

$$v = \frac{(c/\mu)+v_1}{1+cv_1/\mu c^2} = \frac{(c/\mu)+v_1}{1+v_1/\mu c}$$

Since $v_1/\mu c$ is small this can be written:

$$v = \{(c/\mu)+v_1\}(1-v_1/\mu c) = (c/\mu)+v_1-(v_1/\mu^2)-v_1^2/\mu c$$

The last term is the only second-order term (since c is in the denominator) and if this is disregarded one can rewrite as:

$$v = (c/\mu)+v_1\{(\mu^2-1)/\mu^2\}$$

Thus the new velocity is the original, c/μ to which is added a drift velocity which is not the velocity of the water v_1 but is $v_1\{(\mu^2-1)/\mu^2\}$. *This is exactly Fresnel's formula for ether drag.* Thus the relativity theory gives a completely consistent explanation both of the first-order and of the second-order drift velocity effects as observed interferometrically.

CHAPTER 9

INTERFEROMETERS RELATED TO MICHELSON'S

Michelson's Variants

A number of important instruments have been developed which are closely allied to the original Michelson interferometer, and a brief review will be given of some of these, indicating their particular applications without stressing too many details. Michelson was fully aware of many of the potentialities and has described quite a number of variations. Some of these are shown in accompanying diagrams. Fig. 9.1 (a) and (b) show two related variations. In these systems the light, after leaving the beam-

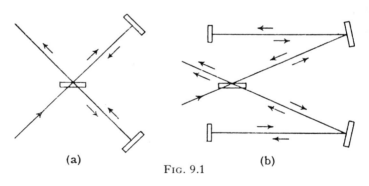

(a)　　　　　(b)

FIG. 9.1

splitter, ultimately strikes a mirror normally and then retraces its path to the beam-splitter. There are, however, a group of variants related to the Mach-Zehnder interferometer, to be described later, in which the beams go round a circuit and meet without retracing their paths. Two such circuits are shown in Fig. 9.2 (a) and (b). In Fig. 9.2 (a) A and B are semi-silvered and C and D are fully silvered mirrors. In Fig. 9.2 (b) there is only one semi-silvered mirror, A. Each of these variations has found its special application.

The Twyman-Green Modification

In 1916 Twyman and Green, by a slight modification in optical set-up, adapted the Michelson interferometer to the testing of prisms and lenses and this has since become recognized as one of the most valuable ways of testing such optical components. It is in widespread use. The essential feature is the introduction of

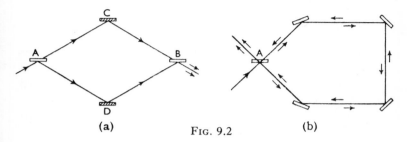

(a)   Fig. 9.2   (b)

strict Fizeau-type collimation. Michelson's set-up normally requires an extended source, but in the Twyman-Green system a point source S at the focus of a well-corrected lens A gives a parallel beam and the arrangement becomes that in Fig. 9.3 (a).

The emerging light is collected by the lens B and the interference pattern is seen by the eye placed at the focal plane D.

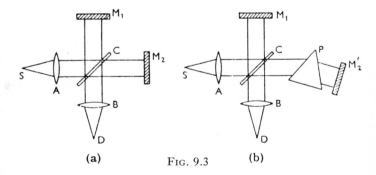

(a)   Fig. 9.3   (b)

Effectively the incidence is restricted to the normal. Consider the case where the mirror and its image are parallel. The extended source as used by Michelson, with its range of incident angles, would produce a set of rings, but the collimated point source as used by Twyman-Green leads to a *field of uniform tint*. Exactly as in the familiar Fizeau wedge system, the fringes are now fringes of equal thickness. If there were irregularities on one of the mirrors M_1, M_2, they would lead to a localized contour fringe pattern.

There are no effective irregularities on the high-quality mirrors used, but if there be interposed an optical component which deforms the wave-front there is an equivalent result and the deviations of the wave-front are seen. The way in which a prism is tested is shown in Fig. 9.3 (b).

Monochromatic light is used, thus there are no dispersion troubles. The prism is introduced at P and because of refraction the mirror M_2' requires displacement compared to the standard position. Suppose the prism to be perfect, then the refracted wave will be a plane wave. If the distance CPM_2' is maintained near to CM_1, uniform tint fringes are seen. If the prism is imperfect, then the wave-front is distorted (twice, once at each traverse) and a contour pattern appears.

The imperfections of a prism may be of three kinds: (1) the angles may not be precisely what are expected, but this is in many cases of no great importance, (2) the surfaces may not be plane, (3) there may be local knots or strains in the glass (especially with large prisms) in which there are slight deviations from the mean refractive index. As a rule most prism faces can be made flat enough, and the main residual cause of distortions is the refractive-index variation. It is thus customary to correct the prism by marking on the surface the observed contour pattern and then working-off portions of the glass locally until a plane wave emerges.

It does not seem to have been fully appreciated, even by experts, that this can be a dangerous procedure, particularly if the original deviations are large. For while it is true that in the interferometer the emerging wave-front can be made quite plane, it is often overlooked that *this is only the case for the single wavelength used for testing*. The dispersion of the glass is involved. The removal of material for a thickness of a light wave in the green to produce a plane wave-front is not proportionate in the blue and the deviation is more so in the ultra-violet. It is the writer's experience that some prisms which give most excellent definition in the region of the green mercury line (which is the test-wavelength in optical workshops) show a steady falling off on either side of the green, the deterioration being more marked at the ultra-violet side. These prisms were possibly of material with initial errors corrected in the manner described. If the initial errors are purely surface errors, then, of course, this feature is not involved.

Lens Testing

Of particular value to the optical industry have been the various lens-testing arrangements introduced mainly by Twyman. One arrangement is shown in Fig. 9.4 (a). Here L is the lens to be examined. Since a lens converts a plane wave into a spherical wave a reconversion to planeness is needed. This is secured by placing a spherical convex mirror M such that the centre of curvature is coincident with the focus of the lens L. Light is thus incident normally everywhere on M and the beams return along their own paths. Only a perfect lens would produce a plane wave-front, but in practice there are tolerances. Rayleigh has established that the images produced by optical systems are practically as good as perfect as long as errors do not exceed a quarter of a wavelength. This theoretical tolerance can often

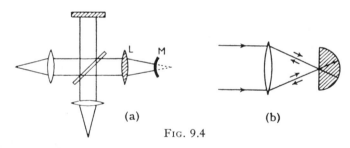

(a)   (b)

Fig. 9.4

be exceeded appreciably in practice. The fringe pattern which emerges from a lens is an integral summary of the various defects (other than chromatic), and an experienced operator can unravel the various effects from the fringe pattern seen and its variation with tilt of the lens.

The lens-testing interferometer is also used for the testing of microscope objectives. This requires a robust accurate construction of the instrumental framework, with fine-adjustment movements. For low-power objectives the system used is that for ordinary lenses, the convex mirror used being a steel ball-bearing of a few millimetres diameter. It is possible to use a drop of mercury for higher powers. A better arrangement, Fig. 9.4 (b), due to Bracey, is for the light to be brought to a focus at the centre of a glass hemisphere the curved face of which is silvered. Any ray entering travels along a radius, strikes the silvered surface normally and returns along its own path.

Köster's Comparator

A number of variants of Michelson's interferometer have been devised for use as length comparators, especially for the standardization of slip-gauges and end-gauges. Such gauges consist of highly polished blocks of corrosion-resisting steel. Köster's instrument is often used for this purpose and has also been used for a precision determination of the metre.

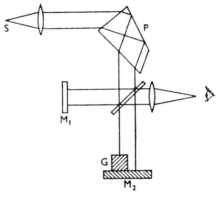

FIG. 9.5

The instrument is in essence Twyman-Green's arrangement, with a prism attached to act as a spectroscopic monochromator. In Fig. 9.5 (a), S is the source, P the prism, on a rotary wavelength-scale turntable, M_1 the movable mirror. The mirror, M_2, is of polished steel and the specimen G whose height is required is "wrung" on to this. Optical wringing is achieved as follows. It is found that plane polished smooth surfaces, lubricated with oil, when pressed together with a twisting motion, bind very closely. It has been established that the binding oil film has a thickness of so small a fraction of a light wave, as to be negligible. The light reflected from the gauge G and the mirror M_2 does not exhibit differences in path due to phase change at reflection since both are polished steel. The field of view, when M_1 is slightly tilted with respect to M_2 is to show two sets of fringes, one passing over the surface of M_2, the other over G. M_1 is usually placed to be half-way between the surfaces of G and M_2 so that the visibility of the fringes from both regions is the same. The fringe patterns show a discontinuity, and the fractional order separation is measured. The source is a Geissler discharge tube, and the fractional orders are measured for a number of selected wavelengths. As will be shown in a later chapter, if the thickness of the gauge is reasonably well known, the measurement of fractions for several wavelengths permits application of a computational procedure known as Benoit's method of exact (or excess) fractions, from which the gauge length can be found to within a small fraction of a light wave. For speed of work special slide-rule methods have been developed for computation of path differences from measured fractional fringe differences for known wavelengths.

The Mach-Zehnder Interferometer

A few years after the invention of Michelson's instrument Mach and Zehnder (1891) independently developed a modification of the Jamin interferometer which has resemblances to some of Michelson's alternatives described earlier in this chapter. The Mach-Zehnder system involving four separate mirrors is difficult to adjust and fell into disuse. During the Second World War it was vigorously applied in Germany to the study of air-flow round aeroplane models, for reasons which will be explained later. There has since been a widespread readoption of the Mach-Zehnder interferometer, using very large apertures, and formidably expensive instruments are in regular use in several countries, particularly for aerodynamic research.

The modern version of the Mach-Zehnder system is shown in Fig. 9.6. This consists of a square or rectangular array with two half-silvered mirrors A_1, A_2 and two full mirrors B_1, B_2. The beams entering at A_1 divide and go round respective halves of the circuit to meet at A_2 and produce interference at C. It is clear that the path lengths in air and glass are identical. Thus, either white-light or monochromatic fringes can readily be secured.

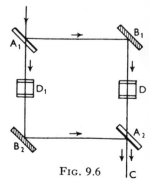

FIG. 9.6

Now suppose the system is especially to be used for the examination of air-flow. The object, a model, is enclosed in a chamber at D in A_2B_1. A compensating chamber with identical windows is at D_1 in A_1B_2. Very high-velocity air is blown past an object in the chamber D (vibration being avoided) and builds up a pressure pattern around the model. The local pressure changes cause local changes in refractive index and thus the air-flow distribution is revealed interferometrically. This is the important application, and for technological studies large apertures and very solid anti-vibration structures are required, apart from the incidental equipment for securing high velocities of air-flow. The glass windows must be very massive to withstand the air blasts without distorting.

There are two reasons why, for such investigations, the Mach-Zehnder optical circuit is superior to the conventional Michelson arrangement. Michelson's interferometer has indeed been used as a refractometer, or as an instrument for measuring thin-film thickness by interposing objects in one path. It is to be noted that the light beam then *traverses such an object in two directions*, i.e. in going *from* the beam-splitter and in returning *to* it. If the object, such as D in Fig. 9.7 above, has considerable local variations in refractive index, there can be some displacement of the beam, and on its return journey the light beam does not pass exactly along the same path as on the way initially from the beam-splitter. This might well lead to fringe confusion, at least to fringe broadening. In the Mach-Zehnder system the *beam passes through the object in one direction only* and this difficulty does not arise.

However, the primary advantage is somewhat more obscure. It concerns fringe localization. It has already been established that with Michelson's interferometer the fringes are localized *on the mirror itself*, i.e. on the position of apparent intersection of the two mirrors. This is of course *not* the position of location of the object, and one is immediately confronted with the very real difficulty that it is not possible to obtain a photograph of both fringes and object simultaneously in focus, unless one stops down camera aperture to an abnormally small value to give an excessive depth of focus. The consequent loss of light then prevents the rapid exposures needed for such experiments.

In contradistinction, the fringe localization with the Mach-Zehnder interferometer is very flexible. When all mirrors are parallel the fringes are at infinity. If mirror A_1 is rotated counterclockwise, then it will be seen that only the *reflected beam* to B_1 is affected whilst the transmitted beam arrives as before at A_2. Thus, if mirror A_2 now be rotated clockwise, the $A_1B_1A_2$ beam is not affected, but the $A_1B_2A_2$ beam can now be brought to meet the former and where they meet becomes the position of fringe localization. If a screen is placed there, fringes are seen on it without the interposition of any lenses in the system whatsoever. In like manner, by rotations in opposite directions, the fringes can be made to appear either in front or behind of A_2.

Thus one can actually place the fringes at will at infinity, or before, or behind the mirror A_2. It is therefore possible to arrange for the regions of localization of object and fringe to be more or less coincident, i.e. to place the fringes at D, and thus secure the crisp photographs required both of the object and of the fringes simultaneously in focus.

Bates Wave-front-shearing Interferometer

Bates (1947) has used the Mach-Zehnder system, with collimated beam from a point source, in a special way for testing the optical perfection of large surfaces. In the Twyman-Green and other Michelson arrangements, if one wishes to test the uniformity of the wave-front from a surface, it is necessary to have an equally large matching surface. Bates arranges to produce interference between two parts of the *same* wave-front, by a shearing wave-front displacement, using the Mach-Zehnder system. This is illustrated in Fig. 9.7.

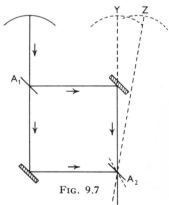

Consider a converging wave-front from a mirror or lens system approaching the beam-splitter A_1. If the four mirrors are all parallel, superposed images appear at Y. If, however, A_2 is tilted, then two images form, Y and Z, and these two wave-fronts can be made to overlap. In a sense Y and Z can

Fig. 9.7

be considered to converge to form two image sources which interfere, the separation being determined by the shear introduced by tilting A_2. These two images can be localized on A_2 itself, and, if this is done, tilting of A_2 will have no effect if the wave-front is truly spherical. Changes in the fringe pattern do, however, appear if there are errors in the original wave-front.

The system can therefore be used to examine the quality of a lens or mirror, in effect by using two images from the same object and superposing one against the other. The use is limited but of considerable advantage in that only the one single object is involved, without a further reference surface.

CHAPTER 10

INTERFERENCE WITH MULTIPLE-BEAMS

Introduction

Interferometry with "*multiple*-beams" instead of with *two* beams has led to very important advances.   By multiple-beams is meant the employ of a succession of coherent beams all in specifically related phase and intensity, these being combined to produce fringes.   The first to recognize the importance of using multiple-beams was Boulouch, who, in a brief note in 1893, gave the theory of formation of a fringe system which was essentially shortly afterwards to be incorporated into both the celebrated versatile

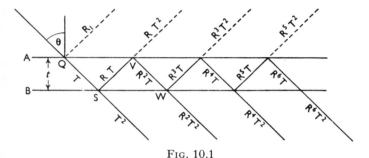

FIG. 10.1

Fabry-Perot interferometer and into the Lummer-plate interferometer.   Boulouch's principal contribution was the recognition of a special property of a formula first derived by Airy in 1836 in connection with the theory of wedge-fringes.   Airy had computed the effect of taking into account the successive multiple images as shown in Fig. 10.1.   These successive images have so far been disregarded because of intensity considerations.   For as will soon be shown the intensity of the nth beam is proportional to R^{2n}, where R is the reflectivity.   For normal incidence on glass R is 0·04, hence even the second beam is very weak compared with the first, and every succeeding beam is so weak as to be of no consequence.

Despite this, Airy made the computation, but this lay dormant for some sixty years.

We shall proceed to derive Airy's formula.   This applies to the cases of both transmission and reflection at two parallel surfaces, the incident beam consisting of parallel monochromatic light.   All the emerging beams are parallel and can be united to interfere at the focus of a lens.   The object of the calculation is to determine the fringe shape and fringe pattern at the lens focus.   For such an arrangement, when using only two beams as already demonstrated, circular fringes (localized at infinity) form at the lens focus.

Airy's Formula

Let us imagine, as in Fig. 10.1, the sequence of reflections and transmissions in a parallel-sided thick plate of glass, bounded by plane parallel faces AB.   The diagram is considerably simplified if refraction in the plate is ignored, for in any event this does not affect the ultimate computation.   At the entrance to the plate, at Q, a fraction of the incident light R_1 is reflected.   This is the *only beam* which does not enter the glass and is the only beam reflected at an air-glass interface.   A fraction T of the incident light is transmitted to S and this, as seen, suffers multiple reflections and transmissions through partial reflection at each event, at a glass-air interface, at which reflectivity is R and transmission T.

We shall assume from the outset that there is no absorption at the interface and therefore for an incident intensity of unity, it follows that $R+T = 1$.   Furthermore, recalling the Stokes Reversibility Principle, this relates the reflectivity amplitudes at glass-air and air-glass interfaces such that $R_1^{1/2} = -R^{1/2}$, a condition affecting the first reflected beam only.   The amount of light T reaching S gives a transmitted beam T^2 and a reflected beam RT.   This latter reaches V where RT^2 is transmitted and R^2T is reflected.   At W, in sequence, R^2T^2 is transmitted and R^3T reflected, and so, indefinitely, provided the plate extends indefinitely.

A series of parallel reflected beams of intensities R_1, RT^2, R^3T^2, R^5T^2, . . ., etc. is created, and a corresponding series of transmitted beams of intensities

$$T^2, R^2T^2, R^4T^2, R^6T^2, \ldots, \text{etc.}$$

Now suppose that the parallel plate is of refractivity μ and thickness t, and let a plane wave (wavelength λ) be incident on the plate at angle θ in the plate (were refractivity to be included then the incident angle would be other than θ). The path difference between consecutive beams in such a plate has already been shown to be $2\,\mu t \cos\theta$ so that each beam lags behind its predecessor by a phase $\delta = (2\pi/\lambda)(2\mu t \cos\theta)$. Since all beams (either reflected or transmitted) emerge parallel, if they are collected with a lens, they will add, and interfere, at the lens focus. Let us represent the incident wave *amplitude* by $\sin\omega\tau$, and compute first the resultant intensity of the reflected light. Remembering that $R_1 = -R^{1/2}$, the amplitudes of the reflected series can be summed as

$$D \sin(\omega\tau - \Delta) = -R^{1/2}\sin\omega\tau + R^{1/2}T.\sin(\omega\tau - \delta) \\ + R^{3/2}T.\sin(\omega\tau - 2\delta), \text{ etc.}$$

Now expand the sine terms and equate coefficients of sine and cosine giving

$$D \cos\Delta = -R^{1/2} + R^{1/2}T(\cos\delta) + R\cos 2\delta + R^2 \cos 3\delta \ldots)$$

$$D \sin\Delta = \qquad R^{1/2}T(\sin\delta + R\sin 2\delta + R^2 \sin 3\delta \ldots)$$

These two combine to give

$$De^{i\Delta} = -R^{1/2} + R^{1/2}T(e^{i\delta} + Re^{i2\delta} + R^2 e^{i3\delta} \ldots)$$

$$= -R^{1/2} + \frac{R^{1/2}T.e^{i\delta}}{1 - Re^{i\delta}}$$

To simplify, multiply and divide the imaginary part by $(1-Re)^{-i\delta}$ making

$$De^{i\Delta} = -R^{1/2} + \frac{R^{1/2}Te^{i\delta}(1 - Re^{-i\delta})}{(1 - Re^{i\delta})(1 - Re^{-i\delta})}$$

$$= -R^{1/2} + \frac{R^{1/2}T(\cos\delta + i\sin\delta - R)}{1 - 2R\cos\delta + R^2}$$

The real and imaginary parts can be separated and equated giving

$$D \cos\Delta = -R^{1/2} + \frac{R^{1/2}T(\cos\delta - R)}{1 - 2R\cos\delta + R^2}$$

$$D \sin\Delta = \frac{R^{1/2}T\sin\delta}{1 - 2R\cos\delta + R^2}$$

Square and add gives

$$D^2 = \left[-R^{1/2} + \frac{R^{1/2}T(\cos\delta - R)}{1 - 2R\cos\delta + R^2} \right]^2$$
$$+ \left[\frac{R^{1/2}T\sin\delta}{1 - 2R\cos\delta + R^2} \right]^2$$

This reduces to

$$D^2 = \frac{4R\sin^2\delta/2}{1 - 2R\cos\delta + R^2}$$
$$= \frac{4R\sin^2\delta/2}{(1-R)^2 + 4R\sin^2\delta/2}$$

This represents Airy's formula for the *reflected* intensity. From this it is easy to obtain the transmitted fringe system (for there is no loss by absorption). Let B^2 be the *transmitted* intensity, then

$$B^2 = 1 - D^2$$
$$= 1 - \frac{4R\sin^2\delta/2}{1 - 2R\cos\delta + R^2} = \frac{(1-R)^2}{1 - 2R\cos\delta + R^2}$$
$$= \frac{(1-R)^2}{(1-R)^2 + 4R\sin^2\delta/2}$$

But $1 - R = T$ if there is *no* absorption, hence

$$B^2 = \frac{T^2}{(1-R)^2 + 4R\sin^2\delta/2}$$
$$= \frac{T^2}{(1-R)^2} \frac{1}{1 + \{4R/(1-R)^2\}\sin^2\delta/2}$$

which is Airy's formula for the *transmitted* intensity.

When $\sin^2\delta/2 = 0$ this is a maximum and has the value $T^2/(1-R)^2$, which is unity since $T = 1-R$, hence *without absorption* the transmitted fringe maxima have intensity equal to that of the incident light, *no matter what the reflecting coefficient may be.* This occurs when

$$2\mu t\cos\theta = n\lambda$$

The minima occur at $2\mu t\cos\theta = (n+\frac{1}{2})\lambda$, i.e. when $\sin^2\delta/2 = 1$. This makes $B^2 = T^2/(1+R)^2$, which can also be written $B^2 = \{(1-R)/(1+R)\}^2$. Clearly when R is near to unity the minima are very weak.

B^2 can be written:

$$B^2 = \frac{1}{1 + \{4R/(1-R)^2\}\sin^2\delta/2}$$

The following is a simple graphical method for deriving the shape of the transmitted fringe distribution given by the Airy formula (reflected fringes, being complementary, are obtained by inverting the curve obtained). Since

$$B^2 = \frac{(1-R)^2}{1-2R\cos\delta+R^2} = \frac{T^2}{1-2R\cos\delta+R^2}$$

construct a triangle within a circle of radius unity as shown in Fig. 10.2 (a) in which $XY = 1$. Let the distance XZ be equal to the reflection coefficient R and let angle $YXZ = \delta$. Then

$$ZY^2 = 1-2R\cos\delta+R^2$$

Clearly as Y sweeps round the circle (i.e. as δ goes through to complete an order) then $1/ZY^2$ will be a measure of the variation of B^2 with δ. Hence a plot of $1/ZY^2$ against δ will give the fringe shape. The intensity values can be obtained by multiplying the ordinates by T^2. Since $XZ = R$ in the limit $\delta = 0$, ZY equals T, thus the maximum value is unity.

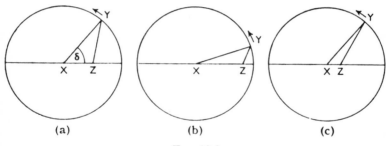

(a)   (b)   (c)

Fig. 10.2

It can easily be seen from this method how the fringe shape depends on R. Take the case where R approaches unity, as illustrated at (b). For small δ value (near $n\pi$) YZ is quite small so that $1/YZ^2$ is large, but, as can be seen, YZ increases at a rapid rate as Y sweeps round, hence the intensity drops rapidly.

But in Fig. 10.2 (c) where R is small, the quantity YZ is initially large and only increases slowly as Y moves round, hence broad maxima result.

The Fringe Width

It is now clear that because maxima appear at $n\lambda = 2\mu t \cos\theta$ the multiple-beam fringes given by a plane parallel plate are simply circles in the identical positions as those already derived for the simpler two-beam case. They are localized at infinity and the ring diameters obey the $\sqrt{n}$ rule. Fractional orders can be evaluated as before. The characteristic difference lies in the fringe *shape and width and these depend only on R.* If we disregard for the moment the variation in separation between orders produced by the $\sqrt{n}$ rule, then the way in which the intensity dis-

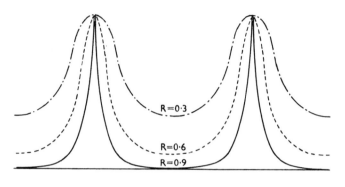

FIG. 10.3

tribution varies with reflectivity is shown in Fig. 10.3 for three arbitrary values of R, namely $R = 0.3$, 0.6, 0.9. Two features characterize each distribution, i.e. the fringe width and the value of the minimum between the fringes.

The fringe half-width is defined as the width at half the intensity maximum and its value is easy to obtain. It is only necessary to find the value of $\delta/2$ in the formula (10.3). This is usually written as:

$$B^2 = \frac{1}{1 + F \sin^2 \delta/2}$$

in which F is $4R/(1-R)^2$ and was called by Fabry the "coefficient of finesse", since, as will be soon evident, it determines the fringe width. In Fig. 10.4 OX, the semi-half width is obtained by putting the half value $B^2/2$ in the intensity expression and finding the value of δ' corresponding to this.

We have: $\qquad \frac{1}{2} = \dfrac{1}{1 + F \sin^2 \delta'/2}$

i.e. $\qquad \sin \delta'/2 = 1/\sqrt{F}$

Since the fringe width is narrow, δ' is small, and $\sin \delta' = \delta'$, hence $\delta' = 2/\sqrt{F}$. But from one order to the next involves a phase change of 2π, hence δ', as a fraction of an order, is $1/\pi\sqrt{F}$. The total half-width XY is, then, $W = 2/\pi\sqrt{F} = 0.63/\sqrt{F}$.

The half-width is thus inversely as the square root of F. But $F = 4R/(1-R)^2$, hence $W = 0.315(1-R)/\sqrt{R}$. It depends therefore only on the reflectivity R. The values of F for $R_1 = 0.3$, $R_2 = 0.6$ and $R_3 = 0.9$ adopted for Fig. 10.3 are respectively 2.5, 15, and 360. The great importance of a high R value becomes evident if we use $R_4 = 0.97$ which is about as high as one can

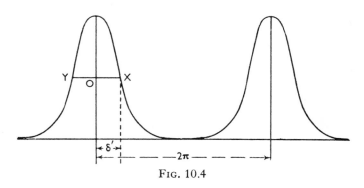

Fig. 10.4

generally get in the green with silver films. This leads to $F = 4300$. Thus an increase in R of *merely* 7% from $R_3 = 0.9$ to $R_4 = 0.97$ increases F in the ratio of over 12:1 and gives a fringe width of less than a third that for the lower reflectivity R_3. Since the power of instruments using fringes is largely determined by fringe width, the importance of a high reflectivity is evident.

The fringe half-widths for the above values of reflectivity R_1, R_2 R_3 and R_4 are respectively 2/5, 1/6, 1/30 and 1/104 of the separation of an order. Thus for the higher reflectivities the fringes are very narrow indeed. The frontispiece (a) shows fringes with $R = 0.90$.

The Minima between Orders

The intensities of the minima, which are half-way between the orders, are given by $[(1-R)/(1+R)]^2$. It is important to note that this never goes down to zero, i.e. that there is *always* some light in the minima. The value of the minima is significant only in the case when interference fringes are being used for spectroscopic studies, for it then sets a limit to the intensities of the weakest lines that can be detected. The sharper the fringes the lower is the minimum and therefore the more favourable the arrangement. For the four above reflectivities R_1, R_2, R_3 and R_4 the minima have percentage intensity values respectively 29%, 4%, 0·3% and 0·02% that of the maxima. At the same time it will be recalled that the transmitted *maxima* all have unit intensity, irrespective of the reflectivity coefficient. Thus even if 97% of the light is reflected away at the first surface the transmitted peaks have the same intensity as the incident light. The integrated quantity of light in the fringes is of course in this example only 3% of the total incident energy.

The Reflected System

It is clear both from the formulae and from energy considerations that the reflected system must be exactly complementary to the transmitted system, for $B^2 = 1 - D^2$. Thus the fringe intensity distribution in reflection is obtained merely by inverting the curves of Fig. 10.3, as shown schematically in Fig. 10.5 for one fringe type.

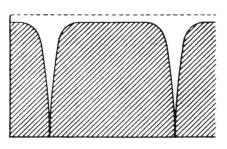

FIG. 10.5

The fringes are now narrow dark fringes on a bright background. They differ in two characteristics from the transmission fringes. First, is the fact that the minima do now really go down to zero (true only when there is no absorption). Second, the maxima are not quite unity but are that distance below unity represented by the height of the minima in the transmitted system. As the latter is $\{(1-R)/(1+R)\}^2$ the maxima here are $1 - \{(1-R)/(1+R)\}^2$ which simplifies to $4R/(1+R)^2$. For fairly large R values this is always nearly unity. For $R_3 = 0·9$ it equals $0\cdot997$.   Even for $R_2 = 0\cdot6$ it still equals $0\cdot96$.   Effectively therefore the maxima can be treated as unity, irrespective of the reflecting coefficient, and the minima as zero.   In practice this condition breaks down, as we shall see, because of the absorption due to the silver film.

The Effect of Absorption in the Silver Film

Whilst it has long been recognized that any absorption in the silver film would necessarily reduce the *intensity* of the transmitted fringes, it was not until 1946 that Tolansky showed that absorption has a much more profound effect on the reflection fringes and can have disastrous results if it is too large.   For *transmission* fringes the effect of absorption is easy to calculate.   Instead of having $T+R = 1$, suppose a fraction A is absorbed by the silver film and thus lost.   Then $A+T+R = 1$.   We had, formerly,

$$B^2 = \frac{T^2}{(1-R)^2} \frac{1}{1+F \sin^2 \delta/2}$$

for the Airy formula, and postulated $1-R = T$, giving $T/(1-R) = 1$.   Now instead we must write $T/(1-R) = T/(T+A)$ (for $1-R = T+A$).

The effect of absorption A is thus to reduce the intensity of the whole overall pattern in the ratio $\{T/(T+A)\}^2$, i.e. by

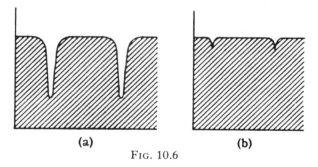

(a)   (b)

Fig. 10.6

$\{1/(1+A/T)\}^2$.   What is significant then is the ratio of absorption to transmission, i.e. A/T.   Now as the reflectivity of a silver film increases, so does A and this can very seriously reduce the transmitted intensity, even by a *factor* of 50 in extreme cases !   *The fringe shape is not affected, the whole distribution being proportionately reduced throughout.*

In the reflected system the calculation of the effect is somewhat more complex, yet in a general manner what happens can be seen this way. In Fig. 10.1 we see that the absorption in the silver film does not affect the first (important) beam of intensity R_1. It does affect every succeeding beam. The intensity of the second reflected beam is RT^2, but the transmission is no longer $(1-R)$ but is reduced to $(1-R-A)$ where A is the absorption. The result of this is that the sequence of reflected beams, which produce the dark interference fringe, is weakened, and as a consequence the fringe minimum does not go down to zero. The greater the absorption the more is the fringe affected. The effect is shown schematically in Fig. 10.6. In Fig. 10.6 (a), with moderately low reflectivity and low absorption the fringes are broad but visibility is fairly good, i.e. the fringes dip down low. In Fig. 10.6 (b) reflectivity is high, with sharp fringes, but absorption increases with higher reflectivities, consequently the fringe dip is only slight and visibility is poor. The fringes are narrow but hard to see. One can imagine the sharp narrow reflection fringes as being diluted with a continuum, the higher the absorption the more the dilution.

CHAPTER 11

THE FABRY-PEROT INTERFEROMETER

Introduction

It was Fabry and Perot who, in 1897, realized the full implication of Boulouch's proposal, and by developing both the theory and practice of their celebrated plane parallel-plate interferometer. introduced a new era of precision into optical interferometry. The Fabry-Perot interferometer is without doubt the most versatile of all interferometers. This instrument has been used for (1) absolute wavelength measurements giving the highest precision, (2) resolution of spectroscopic line structure (hyperfine structure) or line widths, (3) evaluation of the metre, (4) determination of refractive indices of gases, (5) measurement of small displacements, etc. It is simple, elegant and powerful. We shall consider it here first as a spectroscopic instrument for high resolution. The instrument consists of two thick plane parallel plates of glass (or quartz). The plates may be from 1 to 7 cm in diameter and perhaps from 3 to10 mm thick; the larger the diameter, the thicker are the plates. This is because the surfaces involving interference must be polished as plane as possible. A flatness of

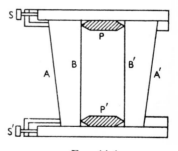

Fig. 11.1

better than $\lambda/50$, i.e. 100 Å, is not uncommon and this can only be secured mechanically with a sufficiently robust plate. Indeed, the preparation of plane discs suitable for a *good* Fabry-Perot interferometer is a peak performance in the optical figuring of surfaces. Each of the two discs is not a plane parallel plate, for the back faces (the unimportant faces) A, A′ are wedged at perhaps 10 minutes of arc, away from the silvered planes B, B′ (see Fig. 11.1). The wedges are sometimes set so that only B, B′ are parallel and no false fringe patterns are produced through

A, A' being parallel. The slight prism formed is of little consequence. The two plates can be variously mounted in a framework
and separated by invar metal spacers p, p', made as accurately
equal as possible. It is not difficult to make p, p' equal to within
say λ/5. With S, S' spring pressure can be applied to bring the
two plates into parallelism to within a very small fraction of a
light wave.

The instrument is used in practice with a variety of spacers
p, p' which may range in value from 1 to 200 mm. It is unusual
to exceed this range, for a variety of reasons. An instrument
with a fixed spacer is frequently called an "etalon".

The Reflecting Films

The selection of suitable reflecting films involves two distinct
matters namely: (a) the intrinsic reflectivity, particularly as a
function of wavelength, (b) the amount of transmission and
absorption shown by a thin film and the way in which this is
related to reflectivity. After much experimenting two types of
reflecting films have more or less universally been adopted. These
are silver and aluminium. The reflectivity of massive silver is

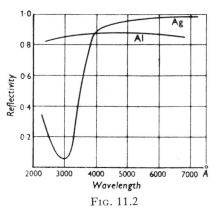

FIG. 11.2

about 0·97 in the red and falls to about 0·90 at 4000 Å. It then
drops very rapidly to merely 0·04 at about 3100 Å. On the other
hand, aluminium has a much more uniform reflectivity. Fig. 11.2
shows the two curves. For the visible region of the spectrum
silver is the best metal, but below 4000 Å silver is of little use and
aluminium is widely adopted. In the near ultra-violet, alloys
of magnesium and aluminium are very useful.

However, the reflectivity of the massive metal is only one aspect. Equally significant is the way in which the reflectivity and transmission depend on the film thickness. The metal films used for interferometers are now most frequently produced by thermal evaporation. The optical plates are placed in a bell-jar which is evacuated down to a pressure of 10^{-5} mm Hg or less. The metal (silver or aluminium) is heated on a tungsten filament, and at the low pressure the mean free path exceeds the distance between filament and optical flat (usually some 30 cm). In a matter of minutes it is possible to deposit a high-reflecting mirror of extreme uniformity, often uniform down to molecular dimensions. Controlled thickness of film can readily be achieved. The optical properties depend critically on the thickness. The reflectivity

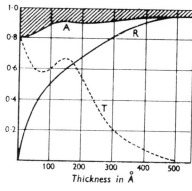

FIG. 11.3

increases and the transmission falls as the thickness grows. At a thickness of about 500 Å (one-tenth of a light wave) reflectivity has nearly reached its maximum value, and beyond this there is a rapid increase in absorption with reduced transmission and very little gain in reflectivity. Fig. 11.3 shows schematically what is found for silver with green light in which R, T, A are the reflectivity, transmission and absorption curves. At 500Å, with a good film, one has perhaps $R = 0.94$, $T = 0.01$ and $A = 0.05$.

For aluminium, absorptions are about double this amount. Now it will be recalled that transmitted fringes are reduced in intensity in the ratio $\{1/(1+A/T)\}^2$, and since for silver A/T may be 5 for $R = 0.94$; this means a reduction in intensity by a factor of no less than 36. This can be a serious matter especially for weak lines.

Special vacuum techniques have been developed for producing optical metal films and some workers have reported that absorption of as low as 0·025 can be secured for high-reflecting films, but these are apparently rare and are usually not always reproducible. They rise rapidly to higher values when the film is exposed to air. Within recent years the use of multilayer reflectors has been advocated. With these, reflectivities as high as 0·99 with absorption as low as 0·01 are attainable. Such a reflector is highly efficient, but only in a very restricted wavelength band and this has limited application.

Use of the Interferometer for Hyperfine-structure Examination

The interferometer is widely used for the measurement of the structures of spectrum lines, the so-called hyperfine structure. Very many spectral lines are found, when examined with high resolution, to consist of close groups of component lines.

This hyperfine structure can have two possible origins. In the first place the spectral source may consist of an atomic species which has several isotopes. The different isotopes have different masses and consequently slightly different nuclear volumes. As a result of either the simple direct mass effect, or the nuclear volume, the lines from each isotope can be slightly displaced one from the other, leading to what is called isotopic hyperfine structure. Then again there is another possibility. It has been established that all odd-atomic-weight isotopes (and some even atomic weights) possess a nuclear spin. The spinning charged nucleus produces a small nuclear magnetic field which couples with the magnetic field of the electron spinning and rotating in its orbit. As a result some spectral terms split up, leading to a hyperfine structure in certain lines. The analysis of such structures is of considerable importance since one can obtain from them: (a) nuclear spin, (b) nuclear magnetic moment, (c) nuclear electric quadrupole moment (another way of describing deviations in nuclear shape from spherical symmetry), (d) isotopic volume changes and (e) isotope abundances.

By far the more important work in this field has been carried out with a variable-gap Fabry-Perot interferometer, i.e. one in which a number of discrete plate separations can be used. Such an instrument has the property of resolving spectroscopic lines which are very close to each other in wavelength, according to the following.

Consider the fringes given by a single wavelength λ. This produces rings at angular positions θ given by $n\lambda = 2t \cos \theta$. Let the order of interference at the *centre* of the system ($\cos \theta = 1$) be n_0, then

$$n_0\lambda = 2t$$

i.e.
$$n = n_0 \cos \theta$$

If there are two lines close to each other of wavelength λ and $\lambda - d\lambda$, the order of interference at the centre differs for the two. It is simpler to make calculations using the "wave number" instead of wavelength. Wave number ν is the reciprocal of wavelength, i.e. $\nu\lambda = 1$. It is given in reciprocal units of length, i.e. cm^{-1}. Thus the two lines in question have wave numbers ν and $\nu + d\nu$. The formula $n\lambda = 2t$ can now be rewritten as $n = 2\nu t$. Differentiate this to find the way n varies with ν, giving $d\nu = dn/2t$.

If then we can evaluate dn the fraction of an order separation between the two lines, then the wave-number separation $d\nu$ is directly obtainable, if we know t, even only approximately.

This simple formula gives the "spectral range" between orders. For if we write $dn = 1$, then the corresponding resulting wave-number separation (write it $\Delta\nu$) is $\Delta\nu = 1/2t$. This shows that if we have two wavelengths separated exactly by one order, the wave-number separation between these orders is $1/2t$ and this *is uniform for all wavelengths*. The larger the separation t, the smaller is the spectral range. For t equal to 1 mm the spectral range is $5\ cm^{-1}$. For $t = 200$ mm it is $0\cdot025\ cm^{-1}$. These figures may be compared with the wave-number separation of the D lines of sodium, which is about $18\ cm^{-1}$.

Suppose then we have a Fabry-Perot interferometer of plate separation 5 mm, the range is $1\ cm^{-1}$, and if we have two wavelengths separated by one-tenth of an order (i.e. $dn = 0\cdot1$), then the lines are separated by $0\cdot1\ cm^{-1}$. There are several methods for finding dn. One way is to find the fractional orders at the centre ϵ_1 and ϵ_2, as in Chapter 7, then $dn = \epsilon_1 - \epsilon_2$. Another way is to approximate by a linear interpolation (for the two ring systems are separated), and a linear interpolation on rings away from the centre is satisfactory. The way in which ring systems are separated is shown for a hyperfine structure doublet in the frontispiece (b).

One great advantage of the variable-gap interferometer is the unambiguous allocation or orders. For suppose we see a pattern in which there are two components, say a weak and a strong, separated by one-fifth of an order at a particular value of t, as in Fig. 11.4. We do not know whether component "a" belongs to A or to B, i.e. is the displacement one-fifth of an order *outwards* or four-fifths of an order *inwards*. Indeed each allocation may be in error by an unknown additional whole number of integers. For example even b or c, etc., might belong to A.

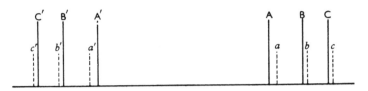

FIG. 11.4

Such ambiguities are resolved by the use of variable separation. For suppose the true structure is a doublet separation of 0.5 cm^{-1}. If $t = 1$ mm, then dn will be just one-tenth of an order. But at $t = 2$ mm this separation occupies one-fifth of an order and so on. Clearly by starting with t small, the order separation Aa is very close, and as t increases Aa widens out. The component "a" superposes on B (and b on C, etc.) for $t = 25$ mm and after this it crosses over and passes B so correct allocation is inevitable.

The important question now arises as to what is the smallest fraction of an order than can be separated, and it is immediately clear that this is a function essentially of the reflectivity in so far as this affects fringe shape. We shall now calculate the resolving power of the Fabry-Perot interferometer.

Resolving Power

The Rayleigh criterion for resolution for slit-images cannot strictly be applied in so far as the fringe pattern now being considered has no secondary maxima and minima. Since, however, the Rayleigh slit-image criterion implies that the diffraction images overlap at intensities which are $4/\pi^2$ ($= 0.405$) of the maximum, it will be reasonably satisfactory to adopt such a value for Fabry-Perot resolution. The calculation is similar to that used for deriving the half-width.

In Fig. 11.5 let the two fringes be just resolved. This happens when the height of B is 0.405 that of the maxima. It is necessary only to find δ_1 ($=$ AB) as a fraction of an order for the point at which the intensity has dropped to 0·405 of the maximum. Since the intensity I at any point is given by:

$$I = \frac{I_{max}}{1 + F \sin^2 \delta/2}$$

then

$$1 + F \sin^2 \delta_1/2 = \frac{I_{max}}{I} = 1/0·405$$

And as δ_1 is small this gives:

$$\delta_1 = 2·42/F^{1/2}$$

The phase separation AC, being twice AB, is then $4·84/F^{1/2}$. Since the whole order corresponds to a phase difference 2π, then AC, as a fraction of an order, is $dn = 4·84/2\pi F^{1/2}$. *The fraction of an order resolvable depends therefore only on the reflectivity.*

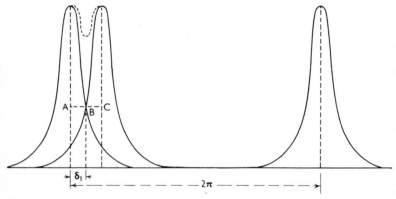

FIG. 11.5

Referring back to our earlier reflectivities $R_2 = 0·6$, $R_3 = 0·9$ and $R_4 = 0·97$, the corresponding F values give the following fractions resolvable. For R_2 it is only about one-fifth of an order. For R_3 it is about one-twenty-fifth of an order, whilst for R_4 it is one-ninetieth of an order.

As separation between orders is in wave numbers $1/2t$, the wave number resolvable is $dn/2t$. For a 1-cm gap and reflectivity $R_4 = 0·97$ this quantity is $0·005$ cm^{-1}. At a wavelength of 5000 Å this is only $0·001$ Å. It is of interest to put this in terms of the classical Rayleigh chromatic resolving power for a spectroscope. This is $\lambda/d\lambda$, and as $\nu\lambda = 1$ then $-\lambda/d\lambda = \nu/d\nu$. For wavelength $5·10^{-5}$ cm (5000 Å), then $\nu = 1/\lambda = 20\,000$ cm^{-1}, thus the resolving power for $R = 0.97$ is $20\,000/0.005 = 4 \times 10^6$. Thus should be compared with the value 10^5 obtainable in the first order of the very best of the largest line gratings.   In theory the resolving power of the interferometer is *unlimited* in that it is proportional to the gap t.   Optical deficiencies are such that the theoretical resolving power is never quite attained, nevertheless resolving powers of 20 millions have often been usefully employed.

Actually the restriction on the use of these higher resolutions lies not in the instruments but in the *widths of the spectrum lines*. Only with very special sources can even 1 million resolving power be employed.   With ordinary arcs and sparks line-width is such as to fill a whole order for even moderately small values of t.

Since the classical resolving power involves λ, it varies across the spectrum.   On the other hand the value $\Delta\nu$, which is the quantity of practical interest to the spectroscopist, remains constant over the whole spectral range (if R does not change).   Thus the true separating power of energy levels is constant, and this is a useful advantage of the instrument.

At this point it is of interest to draw attention to the fact that Michelson's definition of fringe visibility is of little value in classifying the real usefulness of fringes.   For it will be surely conceded that the multiple-beam Fabry-Perot fringes are much superior from the viewpoint of resolution in terms of fractions of an order than are the corresponding two-beam fringes.   Yet because in two-beam fringes the minima in the reflected system *do go down to zero*, the visibility is *perfect*, i.e. it is equal to unity. Even in the best of Fabry-Perot fringes the *minima are larger than zero*, so that the strict application of the visibility concept leads to the somewhat anomalous position that the visibility of two-beam fringes is superior to that of multiple-beam fringes.   Clearly, visibility is not a very useful criterion for classifying fringe properties, half-width is far better.

Use on a Many-lined Spectrum

So far the instrument has been imagined to be used with one wavelength (or with a single close hyperfine structure group). In special cases, e.g. the green mercury line, filters are available for isolating one wavelength.   In general the instrument may be used with a complex line spectrum.   The rings from each line would overlap and confuse if some auxiliary sorting were not carried out. This is achieved by "crossing" the interferometer with a spectrograph, often merely a low-power prism spectrograph. There are several ways in which this can be done, a convenient method being illustrated in Fig. 11.6 (a). This has a high light efficiency. Light from the source S_1 is rendered parallel by the lens L_1 and is incident on the interferometer FP. The well-corrected achromatic lens L_2 throws an image of all the different fringes on the slit of a spectrograph S_2. This is opened very wide, and as a result one has on the photographic plate broad

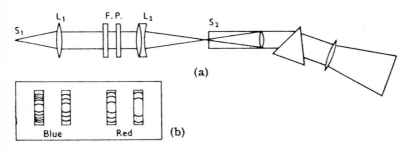

(a)

(b)

Fig. 11.6

line images crossed by the fringe patterns. The spectrograph sorts out the lines, as in Fig. 11.6 (b), and each is crossed with a diametrical section of its circular rings, the section being selected by the wide slit.

By this method one can simultaneously photograph fringe patterns for all the lines of a complex line spectrum. These can be used either for the evaluation of hyperfine structure or for absolute wavelength determinations in the manner described in the next section.

Determination of Wavelengths

By means of a clever device due to Benoit (1898) and known variously as the method of "excess" fraction or "exact" fractions it is possible to derive the wavelengths of lines with very high accuracy, in terms of that of the standard wavelength, red cadmium. It is necessary to have at least three high-precision wavelengths in addition to those for the lines under study, all needing to be simultaneously recorded by an arrangement of the kind shown in Fig. 11.6. It is also necessary to know the approximate wavelengths of all the lines whose precision wavelengths are being sought to at least 0·1 Å.   Yet the procedure is such that one obtains all the wavelengths to within an accuracy of the same order as that of the standard (to perhaps 1 in 5 millions), i.e. to within an error not exceeding 0·001 Å.   Indeed in most recent work an accuracy of 0·0001 Å is achieved.

We shall select as an example the original data given by Benoit. At least three standard lines are required, and Benoit selected the following four lines of cadmium whose wavelengths have been directly measured against the standard metre, with Michelson's interferometer, 6438·4696 (red), 5085·8220 (green), 4799·9088 (blue) and 4678·150 (violet).

Since an air interferometer is used, strictly speaking the wavelengths obtained will be air wavelengths.   If these are to be converted to vacuum wavelengths, then one must use $n\lambda = 2\mu t$, and μ for air is 1·000 293.   One can either evacuate the whole interferometer system, or alternatively correct for the very small dispersion of air (only some 6 units variation in the last decimal place over the whole visible region).

The interferometer used by Benoit had a spacer t, which when measured by a precision micrometer was found to be 10·005 mm. It is necessary to know this to within 1 part in 5 millions, and this information is secured thus.   The fractional orders at the centres are determined for the rings of each of the lines.   They were measured to within perhaps 2% and were given as:

Red	Green	Blue	Violet
0·82	0·00	0·79	0·93

(Nowadays a higher precision obtains.)

Now, if we take $n\lambda = 2t$ and divide the measured $2t$ (20·010 mm) by the known wavelength of the red line (6438·4696), we shall get as an approximate order of interference for the red line 31 050.   It is known in fact from the measurements that the fractional order at the centre for the red line *must be 0·82*.   Clearly the real order must be 31 050·82±x, where x is a small whole number and represents the uncertainty due to not knowing t correctly by mechanical measurement.

Any one of these successive order values will give a corresponding t value, *one of which will be correct*.   One therefore takes these values t_1, t_2, t_3, . . . , etc., and for each of these works out the order of interference for the remaining three lines. A selection from these is shown thus:

Red	Green	Blue	Violet
31 049·82	39 307·97	41 649·40	42 733·41
50·82	09·23	50·74	34·79
51·82	10·50	52·08	36·17
52·82	11·76	53·42	37·54
53·82*	**13·03***	**54·77***	**38·92***
54·82	14·30	56·11	40·30
55·82	15·56	57·45	41·67

It is seen from this table that only for the red order 31 053·82 do the calculated fractional values come out for all the other lines of the same value as those observed (within limits of observation). This then must be the correct order of interference. This permits the absolute spacing to be evaluated by simply multiplying n and λ.

Now that the exact separation is known, all that is needed for wavelengths under investigation is that their orders should be known to within a complete whole number. Taking any unknown line near the 5000 Å, the order of interference is some 40 000. This must be known to 1 unit, hence the wavelength must be known to about 0·1 Å. The experimental fraction is evaluated for this line, and if it is known to say 1 unit in the last decimal place, i.e. to 0·01, then we now know the real order of interference for this line to this accuracy, i.e. 1 part in 4 millions. Thus by starting with a wavelength known to 0·1 Å, we end up with the wavelength known almost to 0·002 Å. This procedure can be extended to a whole complex spectrum. It is the recognized method for evaluating secondary standards of wavelength, the primary cadmium line standards having been evaluated by direct comparison with the metre. With modern techniques, fractions can be more accurately measured than formerly and wavelengths can be intercompared to ±0·0001 Å.

Phase-change Correction

In using the silvered Fabry-Perot interferometer over a wide wavelength range a difficulty arises when high precision is aimed at. This is the fact that when light is reflected at normal or approximately normal incidence on a thin silver film it suffers a phase change, depending on the film thickness and the wavelength. Thus the *optical* separation and the *metrical* separation of two silvered plates differ slightly. Now this difference is dependent on wavelength. For typical Fabry-Perot mirrors it can cause a change which differs from the red end of the spectrum to the blue end by perhaps 0·02 of a wavelength of light.

As the Fabry-Perot interferometer uses the red cadmium line as standard, clearly one can adopt a zero phase error for this wavelength, since all this means is that the plate separation derived optically by the method of exact fractions is the optical, not the metrical separation, at the red cadmium line. This is of no consequence. It is found experimentally that the phase error can be corrected for by subtracting fractions of an order for different wavelengths.

The general method for eliminating the phase error was given by Meggers (1916). One measures the apparent wavelengths with two separate spacers. Let λ' be the standard wavelength at which the phase effect is zero. Any other line λ will have apparent slightly different wavelengths for the two different interferometer spacings, because the adopted spacings are affected by the unknown phase change. Let λ_1 and λ_2 be the measured apparent wavelengths for spacings t_1 and t_2 and let $d\lambda_1$ and $d\lambda_2$ be the phase-change errors arising in the two cases. Then

$$\lambda = \lambda_1 + d\lambda_1 = \lambda_2 + d\lambda_2$$
$$= (2/n_1)(t_1 + \phi) = (2/n_2)(t_2 + \phi)$$

in which n_1 and n_2 are the orders and ϕ the apparent increase in the two paths due to phase.

Since the measured values of λ (i.e. $\lambda_1 \lambda_2$) are expressed in terms of an interferometer thickness derived from the standard λ', then

$$d\lambda_1 = 2\phi/n_1 \quad \text{and} \quad d\lambda_2 = 2\phi/n_2$$

i.e.
$$d\lambda_1/d\lambda_2 = n_2/n_1 = t_2/t_1$$

or
$$d\lambda_1 - (t_2/t_1)d\lambda_2 = 0$$

But
$$d\lambda_1 - d\lambda_2 = \lambda_2 - \lambda_1$$

$$d\lambda_2\{(t_2/t_1) - 1\} = \lambda_2 - \lambda_1$$

i.e.
$$d\lambda_2 = \{t_1/(t_2 - t_1)\}(\lambda_2 - \lambda_1)$$

This is the correction to be added to λ_2 to secure the true value of λ.

Hence from the apparent difference in wavelength, using two interferometer spacings in succession, with of course the same mirrors, the phase effect can be eliminated.

The phase effects for silver, reflectivity near to 0·90, and aluminium, reflectivity near to 0·85, have been determined experimentally by Barrell and Teasdale-Buckell (1950) for visible lines of mercury using a range of interferometer gaps of from 1·6 to 125 mm.

They give for silver the following values of the phase change:

Wavelength	6438	5770	5460	4358	4047
Phase correction	0·000	0·000	+0·003	+0·023	+0·040

For aluminium, the effect over the whole visible spectrum is, on the other hand, negligible (within $\pm 0\cdot002$). The reliability of these phase-change corrections is indicated by the following wavelengths of green and blue mercury lines evaluated with a special single isotope source (see later) with respectively silver and aluminium mirrors:

	Silver	Aluminium
Green	5460·7527	5460·7528
Blue	4358·3371	4358·3372

There is agreement to within 0·0001 Å from the two types of mirror, despite the large effects involved.

Edser-Butler Fringes

When a Fabry-Perot interferometer of very small gap (say less than 0·1 mm) is placed before the slit of a spectroscope and then illuminated with a parallel beam of white light, at normal incidence, a group of evenly spaced parallel fine-line fringes is seen in the spectroscope eyepiece. Such fringes with two beams (in reflection) were described long ago by Fizeau and Foucault, but it was Edser and Butler (1898) who first used silvered surfaces, giving relatively sharp fringes, in transmission. (This fringe system is often called "channelled fringes" in continental texts.)

The characteristics of these fringes are easy to obtain. We have for the Fabry-Perot interferometer $n\lambda = 2\mu t \cos \theta$, and in this special case $\mu = 1$ and $\cos \theta = 1$ so that $n\lambda = 2t$. With white light λ is continuously variable, and in effect for integral values of n there will be fringe maxima. In other words, at those wavelengths $\lambda = 2t/n$ for which n is an integer maxima will occur. This means that only at such wavelengths will any appreciable light be transmitted, hence a system of line images corresponding to these wavelengths, but having the Airy distribution across the wavelength range, will be transmitted.

The separation between the fringes is easy to calculate thus. It is simpler to use "wave number" instead of wavelength. (By definition the wave number ν is the reciprocal of the wavelength, i.e. $1/\lambda = \nu$.)

The formula $n\lambda = 2t$ can be rewritten as $n = 2t\nu$. Differentiating gives $d\nu = dn/2t$, in which $d\nu$ is the wave-number separation corresponding to a change in order dn. In changing from one fringe to the next $dn = 1$, hence $\Delta\nu$, the wave-number separation between orders, is given by $\Delta\nu = 1/2t$ cm^{-1}. This simple expression shows that *the fringes are equally spaced*, in terms of wave number, right across the spectrum. (There is, in fact, a slight variation because of the alteration with wavelength of phase change on reflection at the silver. This can be corrected for.)

If t is 0·1 mm, the wave-number separation is 50 cm^{-1} which is about three times the separation of the sodium D lines. For progressively smaller t values $\Delta\nu$ increases proportionately. If the two surfaces enclose only a thin air film, the separate fringes can be many hundreds, even thousands, of angström units apart.

The uniform separation of these fringes makes them a very useful means for calibrating a spectroscope. They have often been used with particular value in infra-red regions where standards of wavelength are difficult to find.

The Compound Fabry-Perot Interferometer

Houston (1927) first introduced an important principle into Fabry-Perot interferometer spectroscopic practice by employing *two interferometers in series* for hyperfine structure studies. The combination is called a "compound Fabry-Perot interferometer". The two interferometers in series should preferably have plate separations which are simple integral multiples of each other.

This is achieved best in practice (Jackson and Kuhn, 1930) by making the etalons as near mechanically as is possible in the correct ratio. They are then placed in airtight boxes and the air pressure adjusted until exact integral optical ratios obtain. Ratios of from 2:1 up to 10:1 have been employed.

The purpose of the arrangement is to secure high resolution with increased spectral range. The resolution of an interferometer increases proportionate to the plate separation, but, at the same time the spectral range between orders, falls inversely with t. If one has a complex hyperfine structure pattern extending over

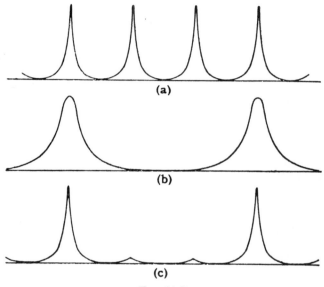

(a)

(b)

(c)

FIG. 11.7

say 1 cm^{-1}, then overlapping of orders, with resultant confusion, occurs if t exceeds 5 mm. But it may be that within this complex there are components so close that they cannot be resolved by a 5-mm gap, yet may be resolved say by a 15-mm gap. If two interferometers are placed in series, one three times the thickness of the other, then one secures the resolving power of the thicker with the spectral range of the thinner. This is made clear by recognizing that a *Fabry-Perot interferometer is an angular filter* only passing light at those angles θ in the expression $n\lambda = 2t \cos \theta$ for which n is integral. Thus in Fig. 11.7 (a) shows the fringe passed by the thicker etalon alone, (b) those passed by the thinner etalon. The fringes (a) are a third of the width of (b) but are separated by only a third of the distance. Clearly the combination passes (c). Here the narrow (and therefore higher resolving) fringes of (a) are passed and now are separated by the distance of (b). One therefore has the resolution of the thicker interferometer, with the spectral range of the thinner. The two intervening fringes belonging to (a) are not quite eliminated. For intensity considerations are such that, because the light has to pass four silver films, it is usually necessary to cut down somewhat on film thickness. The reflectivities are therefore generally not as high as one would like. The lower reflectivity means that the fringe minima are appreciable, hence the intervening maxima are not quite quenched. This can be a serious defect when a hyperfine structure pattern has a wide range of intensities within its components, and in such a case the investigator must be on his guard. In the many important cases where line patterns consist of few components of like intensities, the compound interferometer has proved of very great value.

The theory of the resolution of compound interferometers shows that the intensity distribution of the thinner etalon has a kind of cutting effect on the wings of the intensity distribution of the thicker, such that the finally transmitted fringes are somewhat sharper than that of the original single thick etalon.

For two interferometers of thickness ratio K the resolving limit, i.e. the smallest quantity, resolvable is $1/(1+K)^{1/2}$ that of the thinner. Thus even when both interferometers are the same in thickness, i.e. $K = 1$, there is an improvement in resolution in the ratio $\sqrt{2}$.

CHAPTER 12

MEASUREMENT OF LENGTHS WITH THE FABRY-PEROT INTERFEROMETER

Fringes of Superposition

The Fabry-Perot interferometer has proved to be a most important instrument for the measurement of lengths both small and large. An important technique in this connection is the use of what Fabry has called "fringes of superposition". These are in effect Brewster's fringes, modified by multiple-beams. A thick etalon can give monochromatic fringes, but not white-light fringes because of the large path differences. But suppose there are one behind each other, i.e. in "series", two etalons which have nearly identical thickness, then path compensations occur as shown in Figs. 7.2 and 7.3 and white-light fringes can be formed.

Suppose the two etalons 1 and 2 have thickness t_1 and t_2, a ray suffering a single reflection in 1 and two in 2 can interfere with a ray suffering two reflections in 1 and one reflection in 2. The

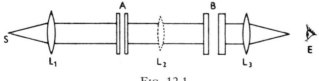

FIG. 12.1

total path difference is $2(t_1-t_2)$ which is zero when $t_1 = t_2$. If it happens that one etalon is not quite parallel sided, i.e. t_2 has a range of values, then one sees a system of coloured fringes across the field when white light is used, but the central *white* fringe occurs where $t_1 = t_2$. This affords a most sensitive way of judging equality of thickness. These fringes are best seen with an arrangement such as Fig. 12.1. S is a white-light point source, rendered parallel with the lens L_1. A and B are the two interference systems. It is advantageous, but not essential, to have an intermediate lens L_2 which is so placed that A and B are at conjugate foci, such that A is imaged on B. The lens L_3 forms an image of the localized fringes which can be seen by the eye at E. These fringes are then *localized* fringes of superposition.

Now it is clear that the same mechanism operates with two etalons one of which is *almost exactly twice* that of the other as in Fig. 12.2. But for zero (or near zero) path difference the beam A makes four reflections in t_1 whilst beam B only makes two reflections in t_2. Thus the two interfering beams *acquire different intensities* (because the reflecting coefficient is less than unity) hence visibility suffers.

Fringes of superposition will arise for any two etalons in thickness ratio n/m in which n and m are integers. The more complex the fraction the poorer is the visibility. It can be shown that the fringe visibility is determined by R^{m+n}. As a rule R cannot be as high as one would wish, since the light has to pass through four absorbing silver films.

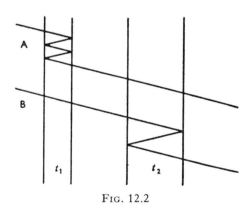

Fig. 12.2

One great mechanical advantage of the use of fringes of superposition is the ease with which a length etalon can be compared with its own length or with simple multiples, without any significant mechanical moving parts being involved.

Fringes of superposition at infinity can also be formed by two *parallel-sided* etalons if these be slightly inclined to one another as in Fig. 12.3. Let A and B be parallel-sided almost identical etalons of thickness t_1 and t_2 inclined to each other at an angle ω. Consider a point Q in the field where PQ makes angle θ with PR which is the perpendicular to the bisector of angle ω. Let this

ray make angles α_1 and α_2 with the normals to the two etalons:
Then the path difference between the interfering rays is:

$$\delta = 2t_1 \cos \alpha_1 - 2t_2 \cos \alpha_2$$

As $\alpha_1 \alpha_2$ and $(t_1 - t_2)$ are all small one can write:

$$\delta = 2(t_1 - t_2) + t_2(\alpha_2{}^2 - \alpha_1{}^2)$$

It can be shown that $\alpha_1{}^2 - \alpha_2{}^2 = 2\theta\omega$.   Thus

$$\delta = 2(t_1 - t_2) + 2t_2\omega\theta$$

Now only θ varies across the field of view as t_1 and t_2 are constant, hence the fringes are *equidistant straight lines parallel to the*

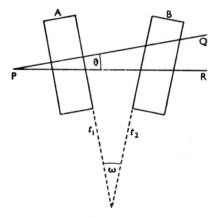

Fig. 12.3

intersection of the etalons.   The angular separation of consecutive
fringes is $\lambda/2t\omega$, where t is a mean value and the central fringe,
for $\delta = 0$, occurs at $\theta = (t_1 - t_2)/t\omega$.

If either t_1 or t_2 be changed (this can be done if the air pressure
in either etalon is altered), the fringes move across the field.   If
the angle of tilt be reduced, the fringe separation increases.

The principle, as before, applies also to two etalons of thickness
in simple ratios, and identical conditions as to visibility hold as
before.

The method can be used to compare precisely two etalons of
near equality.   For t_1 is set normal to the incident light and then
t_2 (the larger) is inclined slowly.   At an angle ϕ, where $t_1 = t_2 \cos \phi$,
there is equality and the white-light fringes appear.   This holds

too for t_1 and t_2 in simple integral ratios. The difference between t_1 and t_2 is therefore evaluated by measuring ϕ. But there is another way in which this can be achieved and this is by using a compensating thin wedge made by silvered flats. Thus let t_2 be nearly $2t_1$ and let t_1 be known. The quantity required is then $t_2 - 2t_1$. The two etalons are placed strictly parallel and illuminated by parallel white-light rays which emerge with the path difference $t_2 - 2t_1$. If this light be now passed through the thin compensating wedge, white-light fringes can form and the central white fringe will appear at that part of the wedge where the wedge thickness is exactly $t_2 - 2t_1$. Since the wedge thickness varies linearly with distance from the edge, a previous calibration permits immediate identification of the difference $t_2 - 2t_1$ between the original etalons.

Evaluation of the Metre

The interferometric evaluation of the metre by Benoit, Fabry and Perot in 1913 using multiple-beam fringes of superposition was a considerable technical advance over the classical determination by Michelson and Benoit in 1892. Mechanically it is much simpler, the time taken is much shorter, only five step-up measurements were used, and the sensitivity is probably a little better. If it were possible to secure fringes over a path length of a metre, then by the method of exact fractions the metre length could be evaluated in virtually a single step. At the time no light source could achieve interference over anything like as long a path. Therefore the experiment aimed at evaluating a 1-metre etalon in terms of an etalon whose thickness could be directly interferometrically evaluated. The "metre" etalon consisted of two silvered plates, separated by a robust invar separator such that they were slightly less than 1 metre apart, actually 99·92 cm. Fine scratches on the upper parts of the interferometer plates define this length. This length is to be interferometrically evaluated and also simultaneously compared with the scratch marks on the standard metre by familiar microscopic methods.

Since a metre is too long for direct interference with monochromatic light (because of line width) the fringes of superposition are used, stepping up by a factor two in each case. The first etalon was 6·25 cm then followed 12·5, 25 and 50 cm. Experimentally the difference in path between successive etalons can be kept down mechanically to less than 40 microns.

Starting with the smallest etalon, the difference between it and the next is evaluated using the fringes of superposition and the silvered wedge compensator in the manner already described. One thus rapidly secures the difference between the "1-metre" etalon and sixteen times the 6·25-cm etalon. The length of the smallest etalon is evaluated directly with a cadmium source, either by the method of exact fractions, or by employing coincidences between the green and red cadmium line fringes. This is rapidly carried out.

There remains for determination the distance between each silvered surface and its scratch mark on the "1-metre" etalon. The sum of the two distances was mechanically as close as possible to 0·8 mm (such that the two scratches were as near as possible 1 metre apart). The plates were removed and made into a thin-gap interferometer whose spacing was interferometrically determined. Then the separation between the scratches (now fairly close) was obtained with a precision-screw comparator. This was repeated for another gap. From the ratio of the two measurements the difference between silvered glass surfaces and scratch marks is obtained in wavelengths.

Sears and Barrell's Determination of the Metre

An elaborate and very accurate determination of the metre was made by Sears and Barrell at the National Physical Laboratory in 1932. They used Fabry's alternative method of compensation, namely that of tilting one of the pair of etalons to be matched instead of using a silvered wedge compensator. Only three steps

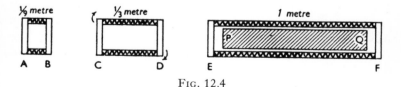

FIG. 12.4

were adopted namely $\frac{1}{9}$, $\frac{1}{3}$ and 1 metre. As before, the interferometric length of the $\frac{1}{9}$-metre etalon was evaluated by the method of exact fractions. The three etalons A-B, C-D and E-F were set up in line as in Fig. 12.4. Of these the A-B and E-F could be displaced sideways whilst the middle one was the only etalon mounted on a tilting frame. This etalon was tilted to match the first and second, the first was removed and then again it was tilted to match the second and third.

A different type of final etalon was employed, namely an end-gauge. This was a steel bar of X cross-section, with highly polished ends P and Q. It was set inside the etalon E-F, and it is the distance E-F which is evaluated by the fringes of super-position. The small distances E-P and Q-F are separately determined by interferometry and thus the distance between P and Q secured in terms of light waves by simple subtraction. (A phase correction allowing for the small fraction of a wave difference between metrical and optical length of the steel bar was predetermined and allowed for.)

The measured end-gauge has now to be compared with the line-standard, which is a bar with lines scratched on. This is done by the use of a composite auxiliary end-bar which consists of a bar $\frac{1}{2}$ inch shorter than the metre, with polished ends. On each end can be wrung a $\frac{1}{2}$-inch polished steel block at the middle of which is a fine scratch. The scratches are nearly 1 metre apart.

A set of comparisons is made between the true line-standard and the composite auxiliary end-bar, reversing the blocks between measurements so that opposite faces contact the bar ends. The average gives the difference between the true line-standard and the auxiliary bar plus half the sum of the blocks. The auxiliary bar, now only with one block at a time wrung on is compared against the end-standard. This is done optically by inserting in turn the end-gauge and the now almost identical composite end-bar (consisting of auxiliary plus one block) into an etalon like E-F in Fig. 12.4. The difference between E-F and the internal bar is interferometrically assessed, so the end-gauge and the composite bar are compared. The average comparison gives the difference between the end-gauge and the auxiliary bar plus half the sum of the blocks. On subtracting, the lengths of auxiliary bar and blocks are eliminated and the difference between the end-gauge and the true line-standard is now obtained. Thus the line-standard has been evaluated.

Special precautions were adopted. The three etalons were built from invar and adjustment for parallelism achieved by straining invar wires. The etalons could be evacuated and thus true vacuum values obtained. The whole equipment was temperature controlled to $0.002°C$. The tilting compensating mechanism for the middle etalon was of the highest precision. Even the correction was made for the expansion of the steel end-gauge when in vacuo, due to removal of the compression of the atmosphere. The change in length due to this cause was no less than -0.22 microns, i.e. about half a light wave.

The Best Values for the Metre

It is unfortunate that the standard metre is a *line-standard*, i.e. is determined by two scratches. If an *end-standard* were adopted (either an etalon of Michelson's type or one of the Sears-Barrell type) a much higher accuracy would result. The cadmium light source itself cannot give interference over more than 50 cm at the most because of its inherent line width, hence the metre has always been evaluated either (a) by successive additions or (b) by optical multiplication. The results can be stated either as the wavelength of the red cadmium line or alternatively as the number of red cadmium waves in the metre. *The International ångström was defined* in 1907 by assigning 6438.4696 Å to the red line (the value then given by Benoit, Fabry and Perot), or, alternatively, as adopted in 1927, there are 1 553 164.13 red cadmium waves in the metre.

Since Michelson's original evaluation in 1892, nine precision determinations have been made, as shown in the following table. Initially, when published, they were not all corrected for standard air conditions, but the figures have now been adapted to dry air, at 15°C and 760 mm pressure, CO_2 normally 0.03%.

1892	Michelson and Benoit	6438.4691 Å
1905	Benoit, Fabry and Perot	.4703
1927	Watanabe and Imaizumi	.4682
1933	Sears and Barrell	.4713
1934	Kösters and Lampe	.4689
1935	Sears and Barrell	.4709
1935	Kösters and Lampe	.4690
1937	Kösters and Lampe	.4700
1940	Romanowa, Varlich, Kartishev and Bartarchukova	.4687

The mean of these is 6438.4696 ∓ 0.0010 Å which, *by chance*, happens to be exactly the value adopted in 1907. The accuracy is 1 part in 6 millions and the greatest departure from the mean is not more than 1 part in 3 millions.

It is of interest to note that *there is no evidence at all of any alteration over the past fifty years* in the length of the standard bar. The optical accuracy is in fact much higher than appears. For law requires that the original prototype must only be handled at rare intervals for the comparison of secondary line-standards. All the experimental work is done against these secondaries. They differ in themselves by 1 part in 3 millions. The average width of the scratches is some 7 μ (i.e. some 10 light waves). Interchanges of *end-standards* made by the National Physical Laboratory, London and the Physikalisch-Technische Reichsanstalt, Berlin, show that optical equality can be established with *end-standards* to 1 part in 30 millions, i.e. five times better than can be achieved with *line-standards*.

The Refractive Index of Air

The refractive index of air was first determined with reasonable precision by Jamin in 1857 using his interferometer, and since that time numerous evaluations have been carried out. In 1939 Barrell and Sears adapted the apparatus with which they had determined the metre to give precision values for the refractive indices of air over the visible range of wavelengths.

In the *wavelength* evaluation of the metre the etalons have lengths in simple integral ratios (3:1). For the *refractive index* studies two etalons are used, as *near equal* as possible. One is evacuated and the other filled with air under controlled conditions. A tilt is introduced to produce fringes of superposition. If now the gas-filled etalon is slowly exhausted the fringes move across the field and the number passing can be measured.

In the experiments the etalon size was such that 700 green-light fringes passed across for a pressure change of 1 atmosphere. Fringe displacement to 0·01 of an order is measurable, giving an accuracy of 1 in 70 000 in the value of $(\mu-1)$. Since $(\mu-1)$ equals 0·000 293 approximately, this gives a final accuracy of 2 parts in 10^{10}. It is clear that the optical accuracy is so great that any errors that will arise will be due mainly to uncertainties in temperature, pressure and composition.

The length of the etalon containing the gas is, as usual, found by the method of exact fractions. A correction is made for the elastic deformation introduced by evacuating the etalons. This was found to be of the nature of 0·6 of a light wave for 1 atmosphere. The actual etalons used had a nominal length of 67·1 cm.

A knowledge of the refractive index in air is necessary to convert precision wavelengths to vacuum values, the relation being $\lambda_{vac} = \mu_{air} \lambda_{air}$.

There exist at least four high-precision determinations of the refractive indices of air and as an example of these, and also as an indication of dispersion values, the following table shows these four refractive indices as reported for both the red cadmium and the green mercury lines. The figures are for normal air, i.e. air at 760 mm, temperature 15°C and 0·03% CO_2.

Wavelength		Meggers and Peters (1918)	Perard (1934)	Kösters and Lampe (1934)	Barrell and Sears (1939)
6438·4896	μ_r	1·000 275 793	1·000 276 410	1·000 276 469	1·000 276 380
5460·743	μ_g	1·000 277 157	1·000 277 896	1·000 277 981	1·000 277 902
	$\dfrac{\mu_r}{\mu_g}$	0·999 998 64	0·999 998 61	0·999 998 49	0·999 998 42

The last row, μ_r/μ_g, shows that from the green to the red the dispersion introduces a change of only $1\frac{1}{2}$ parts per million. There seems little doubt that the mean of these 0·999 998 51 is certainly correct in the second last figure, i.e. to 1 part in 10 millions. Even at $\lambda 4358$ the refractive index has only increased to the value 1·000 281 02, i.e. 5 parts in the sixth decimal place.

The New Single Isotope Source

By taking advantage of a nuclear reaction in the atomic pile a new source has been evolved which produces a whole spectrum of lines, each of which is inherently sharper than the red cadmium line. When a sheet of gold foil (of atomic mass 197) is irradiated in a pile by neutrons, neutron capture leads to the formation of an unstable gold isotope Au^{198}. This is β-active and on emission of a β particle becomes a stable mercury isotope Hg^{198}. Since mercury forms an amalgam with gold, the mercury is not lost. If the gold sheet is irradiated for some weeks with intense neutron beams, on removal of the sheet and distillation in vacuum an appreciable amount of pure Hg^{198} is obtained.

Vacuum discharge tubes can be made containing a few millimetres pressure of argon and as much as 3 milligrams of pure Hg^{198}. Such a tube excited with a high-frequency discharge gives fringes which can be measured over at least a length of 400 mm.

Because there is only one isotope, and as the mass is so much greater than that of cadmium, an improved Doppler width obtains. It is with this source that the high-precision values given in the section dealing with phase-change effects were obtained.

If a 500-mm end-standard can be measured directly, and this is within sight, then the metre is obtainable with either a single optical addition or a simple doubling optical multiplication. This should have considerable advantages in speed and accuracy.

The new source offers other advantages, in that several lines are available, *all* of great sharpness, and this permits the method of excess fractions to be applied with great confidence and speed.

It has been announced by Terrien and Hamon (1954) that interference over a path 1 metre in length has been secured with the infra-red line 9856 Å from a source consisting of a separated single isotope of krypton (mass 86) cooled to a temperature of 63°K. It should be possible with this to measure a metre line-standard with a single operation, but not of course against the red cadmium line.

New Definition of the Metre

Partly as a direct result of the excellence of the interference fringes given by a single-isotope krypton source and partly as a result of other prolonged historical pressures, it was finally decided in October 1960 by the 11th Conference on Weights and Measures to change completely the definition of the metre. From that date on the metre is no longer defined as the distance between specified marks on a specified standard metallic bar at Paris. The metre is now defined in terms of a particular wavelength of light. The wavelength, selected because it gives well-defined fringes over a longer interference path than does the standard red cadmium line, is an orange line emitted by a discharge tube containing as its gas an isolated single isotope of krypton. With modern high-speed isotope separation diffusion techniques the isotope, mass 86, can be extracted from krypton in adequate quantities to enable discharge tubes to be made and distributed.

The selected wavelength is described as the $2p_{10}-5d_5$ transition (using the old term notation for krypton). It is a line of wavelength approximately at 6058 Å.   To be precise the definition now is:

"1 metre equals 1 650 763·73 wavelengths, *in vacuo*, of the krypton line $2p_{10} - 5d_5$".

This value was not obtained by direct measurements made on any standard bar (this would be against the spirit of the definition), but on the contrary it was obtained by a direct evaluation (by five different laboratories), of the wavelength of the line in question matched against the accepted vacuum wavelength of the red cadmium line.   The former standard metre bar was assessed in terms of this red cadmium wavelength in a *standard atmosphere*. From the known accurate refractive index of air this can be converted to a *vacuum* wavelength without increasing the error.   The vacuum wavelength of the accepted red cadmium line becomes then $6440·2490_7$ ångstrom units.   The krypton line wavelength was compared interferometrically with this.   All five separate evaluations agree to within three parts in 10^8.

One additional consequence following from the definition of the metre in terms of a wavelength is that the ångstrom unit now really does become equal to 10^{-10} metre instead of only, as formerly, closely approximating to it.

Thus it has come about that the older (1889) definition of the metre in terms of a metal bar has now been abandoned.   This is an undoubted scientific advantage on two grounds, namely, (i) the bar could slowly recrystallize and creep away (minutely perhaps, nevertheless sufficiently) from its value when first assessed; (ii) it could be subject to accident from fire, flood or explosion and even be destroyed.   Furthermore, even if it is constructed of platinum-irridium, it might conceivably, over long enough, be affected by corrosion.   Nevertheless, although abandoned as the metre standard, it is still agreed that, at the moment, the length of the bar in terms of red cadmium (and thus also in orange krypton) is to be considered exactly what it has been up to now.   It will be for some future redetermination in terms of orange krypton to find out whether the bar changes slowly with time.   Such an evaluation will be easier than with red cadmium because of the better line in the new source.

The krypton source adopted has been carefully specified to enable different laboratories to replicate it closely.   The isotope-86 content must have a purity of at least 99%.   The source is described (in detail) as an Engelhard type hot-cathode glow discharge lamp, which must be cooled to 63°K when under observation, and in the capillary tube defined, it should carry a current of 0·3 A/cm². It is definitely superior to the mercury-198 single isotope source. This source, when operated at a suitable brilliance at a temperature of 0°C, can give observable fringes over a path difference of 50 cm. The new krypton source still shows recognizable fringes over a path difference of 80 cm, being appreciably narrower in line width.

It may be noted that, good as the krypton-86 source is, it may yet be superseded in the future by the still sharper line source known as an atomic beam source. In such an arrangement a narrow directional beam of atoms, in an otherwise very high vacuum, is selected by a pair of slits in line, one well separated from the other. The atoms in the beam travel effectively in one linear direction. They are excited by impact from an electron beam and emission light is viewed in the direction at right-angles to the direction of travel. As a consequence of this, the effective velocity in-line-of-sight is greatly reduced, with the result that the natural Doppler broadening line-width in-line-of-sight is very small. A great improvement in line sharpening results and, indeed, it leads to a Doppler width as narrow as that resulting from cooling down to a few degrees absolute, although no cooling whatsoever is resorted to (Doppler line-width is proportional to the square root of the absolute temperature).

With lines of such a high degree of sharpening it is possible to secure fringes over a length of a metre. It becomes possible then to measure a metre length directly interferometrically without doubling-up procedures of the kind needed with the red cadmium line.

CHAPTER 13

MULTIPLE-BEAM INTERFEROMETRY
I. LOCALIZED FRINGES

The Phase Condition

Although Fabry, Perot and Buisson in 1906 used a thin wedge silvered on both sides, to produce sharpened Fizeau fringes, they did not analyse the necessary conditions required to produce really sharp wedge-fringes. This analysis was first carried out by Tolansky (1946), who pointed out that the Airy summation strictly speaking only holds for a *parallel plate*, but that if certain critical conditions are fulfilled for a doubly silvered wedge, then a close approximation to the Airy summation can be achieved. These conditions are quite critical, but it was established that it is possible to secure very sharp fringes over surfaces, even when high magnifications are employed. This extension of the Airy intensity distribution to all types of Fizeau fringes has proved of considerable value and has enabled a good deal of high-precision work on surface topographies to be carried out. Indeed it has been possible by such means *even to measure many crystal lattice spacings*, with light waves. Furthermore, other types of multiple-beam fringe systems have been developed, once the necessary conditions were recognized. Some of these will be discussed briefly later.

The key distinguishing feature between the wedge and the parallel plate is that in the wedge the successively multiple reflected beams *are not behind each other in phase in exact arithmetic series*. There is a *phase lag* which depends on the thickness of the wedge, the wedge angle and the angle of incidence. This is clear from Fig. 13.1.

Let PQ, RS be the two silvered surfaces, θ being the angle between them. Let the light be incident normally on PQ. Consider interference at the point B produced by the ray AB and the ray CDEB, which is reflected once at each surface. CD turns through angle 2θ to reach E and then through 4θ to reach B. Angle EBA is 2θ. Let AB be t. Then to a first order EA $= 2t\theta$

and CE $= 2t\theta$ so that CA $= 4t\theta$. Hence DB$' = 4t\theta^2$ making CD $= t(1-4\theta^2)$. From the geometry, to a close approximation (as θ is small), ED $= t(1-2\theta^2)$ and EB $= t(1+2\theta^2)$. It follows that the difference between the two paths CDEB—AB is not $2t$ (which is what happens for parallel plates), but $2t-4t\theta^2$.

Suppose now that, instead of two beams arriving at D, the reflecting coefficient is such that successive multiple-beams are of sufficient intensity to influence the resulting interference. Consider, as in Fig. 13.2, a direct beam 1 and two successive multiple-beams 2 and 3. Since the incident beams are parallel and meet at D, they must impinge on the first surface at distances progressively farther from A. The beam 2 is deflected first through

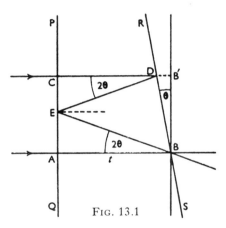

FIG. 13.1

2θ then through 4θ, but the beam 3 is deflected first through 2θ, then successively 4θ, 6θ, 8θ and so on for higher orders.

It is clear that the path difference between successive reflected beams is not constant but alters progressively with the order of reflection. By extending the previous calculation for two beams it can be shown by a calculation (Tolansky) that the path difference between the first and nth beams is $2nt-(4/3)n^3\theta^2t$. Thus the phases of beams of higher order gradually get out of step with the first beam, and instead of assisting the Airy summation series, begin ultimately to oppose it. Indeed, when the retardation is $\lambda/2$, such a beam tends to destroy the condition of sharpness.

This phase condition can be derived by different geometrical methods and the form of the result obtained varies slightly according to whether the light is normally incident on the first or on the second face. Brossell (1947) has simplified the calculation for phase lag and generalized it to include regions other than those on the interference film in accordance with the following Fig. 13.3.

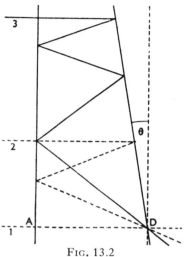

FIG. 13.2

Let AC, CB of Fig. 13.3 represent the wedge and also the wavefronts reflected at each surface. Then $CD_1 \ldots CD_n$ represent the successive wave-fronts after multiple reflections. For the nth beam the angle $D_nCB = 2n\theta$.

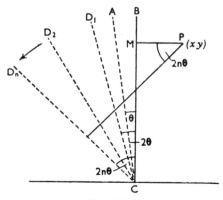

FIG. 13.3

Consider the path differences of the first and nth beams at the point P(x,y).

This is
$$\delta = PN - PM = PN - x$$
$$PN = x \cos 2n\theta + y \sin 2n\theta$$

Thus
$$\delta = x(\cos 2n\theta - 1) + y \sin 2n\theta$$

Expanding the cosine and sine ($2n\theta$ being small) gives:

$$\delta = 2nt\left(\frac{1-2n^2+1}{3}\theta^2\right) - 2xn^2\theta^2$$

By viewing fringes at the surface BC, x is made zero and the retardation lag behind the arithmetical value becomes:

$$(2n/3)(2n^2+1)\theta^2 . t$$

which for the large values of n to be considered, effectively equals $(4/3)n^3\theta^2 t$ precisely what had been derived earlier by Tolansky.

This expression for δ has the added advantage of showing that when x is not zero the quantity $2xn^2\theta^2$ can be such that sharp fringes appear in space in a sequence of positions *other than on the wedge surfaces*. This is a refinement which will not be pursued here.

The Importance of a Small Gap

The lag $(4/3)n^3\theta^2 t$ shows that the Airy sum condition will in general be secured *only when t is very small*, for the following reasons. With the high reflectivities available n can be very large, indeed no less than sixty to eighty effective beams can often be seen by eye to contribute to the sum total effect. Suppose we take n to be 60. Let us consider that the lag is such that by the sixtieth beam it has mounted up to half a wavelength of the light used. Clearly any beams which are half a wave behind will oppose the Airy summation and thus reduce definition. This will partially hold also for a considerable number of beams on either side of the sixtieth. Let us take this retardation as a permitted limit for the sixtieth beam, i.e. equate $(4/3)n^3\theta^2 t_c = \lambda/2$ for $n = 60$, letting this critical value of t be called t_c.

The value θ determines the number of fringes per centimetre across the field of view. Suppose there are X fringes per cm, then $\theta = \lambda X/2$. Thence:

$$t_c = 3/2n^3\lambda X^2$$

Taking $\lambda = 5000$ Å and substituting $n = 60$, gives:

$$t_c = 1/7\cdot2X^2 \text{ cm}$$

This gives an upper limit for t, and, indeed, for the Airy sum condition to apply, t should be appreciably less than this such that even the sixtieth beam is certainly appreciably less than $\lambda/2$ behind.

The value for t_c depends upon the square of the number of fringes per centimetre on the surface, i.e. on fringe dispersion. If dispersion is high, say fringes are 1 cm apart, then t_c must be less than 1·4 mm. If, however, fringes are 1 mm apart, then t_c must be less than 0·014 mm. Now this is only about 28 light waves thus we see that when fringes are nearer than 1 mm apart then t must approach the dimensions of light waves.

In fact small values of t are usually necessary for many applications, for these fringes are particularly useful for the examination of surface topography. It has become increasingly necessary to make observations over smaller and smaller areas, using a microscope. Clearly if the area under review is, say, 0·1 mm across, then in order to have at least two fringes in the field of view the fringes must needs be 0·05 mm apart at most. For four fringes to be seen, a t_c of about 0·001 mm is needed under such conditions.

Linear Displacement of Beams

Apart from the phase retardation condition there are other important factors which are also dependent on t. Referring back to Fig. 13.1 it is seen that the beam CD strikes the interferometer at a distance from A equal to $4t\theta$. The higher order beams in Fig. 13.2 come from progressively farther regions. To a first approximation the linear separation on the surface of the wedge between the first beam and the nth beam is

$$d = 2n(n+1)t\theta$$

which for n large is $2n^2t\theta$.

As $\theta = \lambda X/2$ then $d = n^2t\lambda X$. Taking $n = 60$ and $\lambda = 5000$ Å this leads to $d = 0\cdot18tX$. Thus the displacement of beams is proportional to t.

When examining a surface with complex topography it is of the utmost importance to arrange conditions such that all the effective beams come from a small region. The reason for this

is clear from Fig. 13.4, where it is seen that considerable confusion of beams reaching D will take place if these beams strike
different topographic features P, Q, R, etc. Indeed such arbitrary path variations can easily be far more destructive to good
definition than the phase lag.

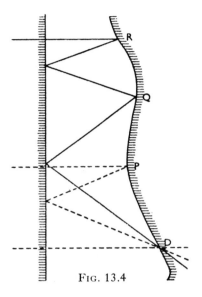

FIG. 13.4

If the critical t_c value is not exceeded, then this displacement
effect is automatically reduced to negligible proportions thus.
Let d_c be the displacement when adopting the gap t_c, then

$$d_c = n^2 t_c \lambda X \text{ cm}, \quad \text{but} \quad t_c = 3/2n^3 \lambda X^2$$

Whence by substitution $d_c = 3/2nX$ cm. Taking as before
$n = 60$ gives $d_c = 1/40X$ cm. For $X = 100$ (fringes 0·1 mm
apart), $d_c = 0.0025$ mm hence all the beams combine from a
region which is about that of the resolving power of a medium-
power microscope. In other words the beams involved in producing a fringe come effectively from what visually appears to
be a point, and there is no beam confusion. It will be clear
that when fringes are produced by two very flat surfaces, the
linear displacement effect does not produce any confusion *per se*
and only the phase lag is significant. This is, however, in practice
a trivial and unimportant case.

PLATE I

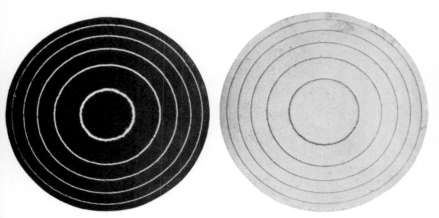

(a) Multiple-beam Newton's rings in transmission.

(b) Multiple-beam Newton's rings in reflection.

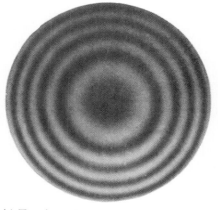

(c) Two-beam Newton's rings in reflection.

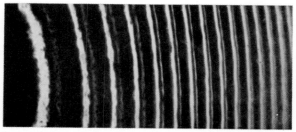

(d) Differential phase-change doubling in multiple-beam Newton's rings. Angle of incidence 50°.

PLATE II

(a) Multiple-beam localized fringes showing microtopography of an octahedron face of a diamond.

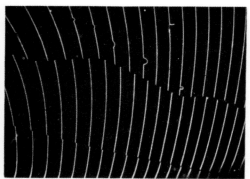

(b) Multiple-beam localized fringes shown by a cleaved sheet of muscovite mica. The discontinuities are the cleavage steps which are often small integral multiples of the crystal lattice spacing.

(c) The "crossed-fringe" system revealing some microtopographic structure on the face of the diamond.

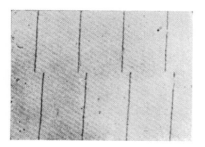

(d) The fringe system applied to the measurement of a thin film, using stepped localized multiple-beam fringes. The discontinuity shown was given by a step of 550 Å. About one-twentieth of this can be measured.

Thus once more it is established that t must be reduced to as low a value as possible. There are two further important advantages of small t which are as follows.

Effects due to Incomplete Parallelism

It has been assumed up to now that a strictly parallel light beam is used. This implies a point source and a perfect lens. Too small a source is not practicable because of intensity considerations. The finite size of the source together with lens imperfections lead to a range of angles of incidence and this produces fringe broadening. At normal incidence the order of interference for a gap t is $n = 2t/\lambda$. For incidence ϕ this diminishes to $n - \delta n = (2t/\lambda)\cos \phi$ in which δn represents the broadening as a change in order.

Thus $\delta n = (2t/\lambda)(1 - \cos \phi) = (2t/\lambda)2 \sin^2 \phi/2$ and as ϕ is small, $\delta n = (t/\lambda)\phi^2$. This change in order, i.e. the broadening is then proportional to t, which must therefore be kept as low as possible. If one agrees to tolerate a certain value δn, then the deviation from the normal, which is permitted, is $\sqrt{(\lambda . \delta n/t)}$, i.e. the permitted deviation is proportional to $1/\sqrt{t}$. For $t = 0.001$ mm this can be as much as $3°$ and this only produces one-two-hundredth of an order broadening. Once again, then, t must be small.

Low Chromatic Resolution

A most important result of having t low is as follows. From the viewpoint of chromatic resolution of different wavelengths a silvered wedge can be considered equivalent to a thin Fabry-Perot interferometer. Now the separation in wave numbers between successive orders in a Fabry-Perot instrument (i.e. the wave-number range) is $\Delta\nu = 1/2t$. If we have an interferometer with $t = 0.001$ mm, this gives $\Delta\nu = 5000$ cm^{-1}. If we have one fringe at wavelength 5000 Å (wave number 20 000), then the fringes on either side will be respectively at wave numbers 15 000 and 25 000 cm^{-1}, i.e. at wavelengths 6666 and 4000 Å respectively. The chromatic order separation is therefore more than 1000 Å. But even when all phase and beam displacement effects are effectively eliminated such that the sharpest possible fringes obtain, reflectivity limitations are such that the best fringes are no narrower than one-fiftieth of an order. This means that the width of a fringe corresponds *chromatically* to something between 20 and 30 Å. This is most important, for it enables broad lines to be used without affecting fringe width. Even a natural line width of 5 Å (which is quite broad) would only be about one-fifth of the width of the sharpest fringe, thus virtually of no consequence.

Since the use of broad lines is permissible *one can use very hot bright sources*. Because bright sources can be used, relatively thick silverings (with high reflectivities) can be employed. Thus, provided t is small the fringe definition can be considerably improved, through permitting high reflectivity. Furthermore, the high intensities enable high-power microscopy to be employed.

Thus, to sum up, the experimental conditions for the production of highly sharpened multiple-beam Fizeau fringes are:

(1) The surfaces must be coated with a highly reflecting film of minimal absorption.

(2) This film must contour the surface exactly and be highly uniform in thickness.

(3) Monochromatic light, or at most a few widely spaced monochromatic wavelengths should be used.

(4) The interfering surfaces must be separated by at most a few wavelengths of light.

(5) A parallel beam should be used (within a 1°–3° tolerance).

(6) The incidence should preferably be normal.

The essential key point, overlooked by earlier workers, is the fact that t_c *must be very small, indeed of the order at most of a few light wavelengths*. Once this is achieved the optical conditions, so to speak, look after themselves.

One is now in the position of securing very sharp multiple-beam localized Fizeau fringes over either large or small areas. All the advantages of precision multiple-beam interferometry can thus be applied not only to the study of large surfaces, but microtopography becomes possible.

The Divergence of the Multiple-beams

Although incident parallel light is used, the successive beams which emerge from a wedge diverge increasingly from the normal. A simple geometrical construction shows that the second, third, fourth, etc., beams emerge successively at angles inclined to the original direction which are 4θ, 6θ, 8θ, etc. The nth beam emerges at angle $\psi = 2n\theta$. As $\theta = \lambda X/2$, then $\psi = n\lambda X$ radian. For $n = 60$ and $\lambda = 5 \times 10^{-5}$ cm this becomes:

$$\psi = 0{\cdot}17X \text{ degrees}$$

For $X = 100$ (fringes 0·1 mm apart), ψ has the surprisingly large value of 17°. That this is indeed the case is readily demonstrated by viewing the fringe pattern with a microscope. On removing the eyepiece the successive beams can be seen straddling right across the aperture of the objective lens.

The illumination is not symmetrical, since the beams displace all in one direction, hence the microscope lens aperture should be such as to be able to collect all the beams in its *semi-angle* of collection. To be fully efficient, then, the total collecting aperture should be 34°. This requires a lens with a numerical aperture of 0·29. If a lens of less aperture is used there is failure to collect all the beams and this leads to broadening of the fringes. The higher the fringe dispersion (the smaller the X) the less need be the collecting aperture.

When all the various optical factors operate, experience has shown that very good definition is still obtainable with microscope magnifications up to × 250, but, beyond that, definition begins to fall off. Yet fringes adequately sharp for many purposes (and still much superior to two-beam fringes) can still be secured at a microscope magnification of × 2500.

Some Applications of Multiple-beam Localized Fringes

Newton's rings. The first application of multiple-beam methods to the modification of long-familiar interference fringe patterns

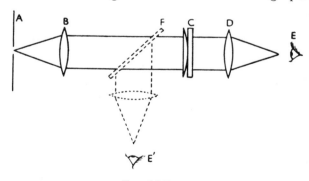

Fig. 13.5

was made on Newton's rings (Tolansky 1944). The optical arrangement is shown in Fig. 13.5. A small aperture A is illuminated with monochromatic light (a bright mercury source with green filter). The lens B gives a parallel beam incident normally on the Newton's rings lens-plate combination C. *The lens and the plate are both coated with a high-reflecting silver layer*, with reflecting coefficient as near 0·95 as is possible, consistent with sufficient transmission of light. *The two surfaces are pressed closely together*, to satisfy the condition of t_c being small. This condition will hold from the centre of the ring system (t extremely small) up to say the twentieth ring, for which t is about 0·005 mm.

The fringes can be viewed with the eye at the focus of a lens D, or alternatively, since they are localized, with a low-power microscope focused on the air film. They can of course be recorded with a camera.

This is the transmitted system, and by interposition of a half-silvered mirror F, the complementary reflected system can be observed at E'.

Photographs of transmitted (a) and of reflected systems (b) are shown in Plate I. It will be seen that the fringes are extremely sharp compared with two-beam Newton's rings shown in I (c) yet they reveal irregularities, which are, in fact, due to the surface microtopography of the polished glass surfaces. For any ring of order n we have $n\lambda = 2t$. An irregularity arises from a local change in t, say dt, and this produces a change in order dn such that $dt = dn(\lambda/2)$. Thus to evaluate the local change in microtopography it is only necessary to measure dn. These sharp fringes have a width some one-fiftieth of an order, and it is possible to detect displacements of one-fifth of a fringe width, hence the limit value of dn which is measurable is one two-hundred-and-fiftieth of an order. This is $\lambda/500$ and for $\lambda = 5000$ Å this is only 10 Å. It is surprising that by such very simple means one can recognize and measure surface details which approach molecular dimensions, *but they are fine details in one dimension only*, i.e. in the up-down direction relative to the surface.

It is emphasized that this very high sensitivity and resolving power *is one-dimensional only*, i.e. in the up-down direction, and *not* across the surface.

Non-normal incidence. When the incident light strikes the system at an angle other than normal a new interference effect appears. This has a bearing on other interferometric procedures and will be briefly surveyed. It is clear from geometry that tilting the lens-plate through an angle will lead to elliptical fringes. Owing to the curvatures, and in accordance with a treatment given earlier, the fringe localization will not be a simple plane.   It will be possible only to secure focus on a plane for a few rings.   The new phenomenon begins to be noticeable when the angle of incidence reaches and exceeds 20°.   The fringes are then found to double.   The doublet separation increases with incidence and the outer fringes move across the field and become progressively weaker and sharper.   The appearance at 50° is shown in Plate I (d) for part of the ring system.   Fringes are in focus only near the centre.

It is found that the two systems of fringes are plane polarized, mutually perpendicularly, *although the incident light is non-polarized.*

The doubling of the fringes arises because when light is reflected from a metal surface at *non-normal incidence,* then, in accordance with familiar classical electromagnetic theory a difference in phase is introduced between light vibrating with the electric vectors respectively parallel or perpendicular to the plane of incidence.   The interferometric gap appears to differ for the two directions of vibration.   If the fringes are separated by a fraction of an order dn, the corresponding difference in path is $dn \cdot \lambda/2$. The path change, as a fraction of a wave, is $dn/2$.   The phase change can be accurately measured by this means, as a function of incidence.   In former classical determinations, the phase change could only be indirectly derived from measurements of the ellipticity of polarization of the light reflected from a metal at non-normal incidence.

The differences both in intensity and in sharpness of the two systems can be accounted for in terms of differential reflectivity and absorption.

At normal incidence differential phase change is zero.   At 50° it is about one-tenth of a wave.   At 20° the doubling is just detectable, but it is well to realize that at about 5° there will still be a broadening effect.   Hence for higher orders of interference with a Fabry-Perot interferometer an improvement in definition can result if plane-polarized light is used.   The improvement will only be slight.

Crystal topography.   Considerable new information has been secured about the surface topographies and growth characteristics of many crystals by the application of localized multiple-beam fringes to these, especially when high-power microscopy is used for viewing the fringes (Tolansky 1945 onwards).   It is necessary to match the crystal with a flat smooth surface.   Observation has established that fire-polished sheet glass, over the small regions required, can be flat and smooth to better than $\lambda/300$, and this suffices as an adequate reference flat for most studies. The fringes can be regarded as forming a contour map, similar to a geographical contour map, but with the separation between each contour line corresponding to a change in height of some $2 \cdot 5 \times 10^{-5}$ cm. Plate II (a) is a typical example of such a complex micro-topography and shows the contour characteristics on a naturally occurring octahedron diamond surface. This is covered with striking triangular flat-bottom or pyramidal depressions.

It is not always easy to tell which regions are elevations and which depressions, but there are several ways of settling this detail, one of which will be described later.

As an example of quite a different type of surface Plate II (b) shows the fringe pattern given by a cleaved surface of a sheet of mica, which is of course a crystalline material. The sharpened fringes show that discrete steps occur on the surface, for there are distinct discontinuities. These each correspond to a distinct cleavage step. Such steps have been measured on many mica samples and are usually found to occur in small integral multiples of 20 Å. i.e. steps of 20, 40, 60, 80 Å, etc., are quite common. Now it is a striking fact that the crystal lattice spacing as found by X-rays is almost exactly 20 Å. Clearly then the steps occur in small integral multiples of the lattice spacing, as indeed is to be expected on crystallographic grounds. But at the same time it is notable that the interference fringes have established this fact and one can alternatively infer from the fringes, with confidence, that the crystal lattice spacing is indeed 20 Å.

Amongst other lattice spacings found by this method are those of selenite, silicon carbide, topaz and several others. The accuracy in the best cases is about ± 1 Å, which is much less than that of X-rays, yet nevertheless it is gratifying indeed to be able to use visible light waves for the evaluation of such small molecular distances. Until the advent of multiple-beam interferometry it was considered that only X-rays and electron beams could measure crystal lattice spacings.

Multiple-beam Interference Contrast. When dispersion is increased so that the field of view includes less than one fringe order, multiple-beam fringes have the property of revealing surface structure with a remarkably enhanced contrast, making visible subtle details *not normally observable*. The principle of this interferometric enhancement of contrast is based upon the advantageous intensity distribution in multiple-beam fringes. Thus, suppose we have two silvered surfaces which are set parallel and are in a monochromatic multiple-beam set-up. Let the two surfaces be slowly moved apart. The amount of light transmitted will vary with distance. This variation is of the identical

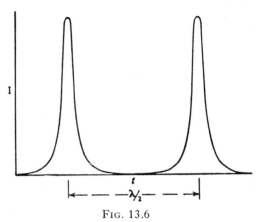

Fig. 13.6

form as the Airy distribution as shown in Fig. 13.6 which gives the plot of transmitted intensity I against plate separation t, successive orders following for increases of t by $\lambda/2$. Now imagine one surface to have on it a discrete step, as in Fig. 13.7, so that there are two distinct interferometers, side by side, of thickness t and t' respectively. The two regions will transmit

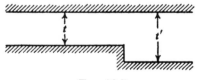

Fig. 13.7

different amounts and therefore will appear as regions of different contrast. The change in intensity produced by a change in t clearly depends upon the particular position of t in Fig. 13.6. If, for example, t is such as to be in the *minimum*, quite a large change in t makes *practically no difference* to the light transmitted. The case is very different if, as shown to a different scale in Fig. 13.8,

t is chosen to be near the maximum. Symmetry considerations indicate that maximum sensitivity will occur if t is selected such that approximately half the peak intensity is being transmitted.

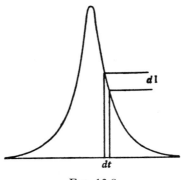

FIG. 13.8

A computation shows that in the neighbourhood of this region, the system is amazingly sensitive to a small change in t. Thus at the half-intensity we have the value of δ given by:

$$\tfrac{1}{2}I = \frac{I}{1+F \sin^2 \delta/2}$$

whence, as it is small

$$\delta = 2/F^{1/2}$$

Now the eye can recognize the difference in contrast when two adjacent regions have a 10% change in intensity. It is simple to evaluate the change in thickness $t'-t$ needed to produce such a 10% change. Let the second region transmit intensity $0.45I$ when the first region transmits $0.50I$ (i.e. 10% change). Then the value δ' for this is given by:

$$0.45I = \frac{I}{1+F \sin^2 \delta'/2}$$

$$\delta' = 2.21/F^{1/2}$$

Hence
$$\delta'-\delta = (2.21-2)(1/F^{1/2})$$
$$= 0.21/F^{1/2}$$

But $\qquad \delta/2 = 2\pi t/\lambda \quad$ and $\quad \delta'/2 = 2\pi t'/\lambda$

calling $\qquad\qquad\qquad t'-t = dt$

then $\qquad\qquad\qquad dt = 0.21\lambda/4\pi F^{1/2}$

which very closely equals $\lambda/60F^{1/2}$.

The sensitivity depends only on F, i.e. on the reflectivity.   For reflectivity 0·94, F is 1044, giving:

$$dt = \lambda/1940$$

For the green mercury line ($\lambda = 5460$ Å) *this quantity is closely enough* $dt = 3$ Å.   This is a remarkable and somewhat surprising result.   For it shows that a change of thickness of *merely* 3 Å (and many crystal lattice spacings exceed this) can cause a recognizable change in transmitted intensity.

Perhaps a more noteworthy way of indicating the extreme sensitivity of this optical system is the fact that a *single lattice spacing in mica* (20 Å) makes no less than a 50% alteration in transmitted intensity !   It has been established that on special types of muscovite mica, when cleaved and silvered on both sides, regions extending occasionally over some square *centimetres* exhibit perfect uniformity of transmitted intensity.   These regions must have *cleaved true to a single molecular layer* over these large areas, and it is thanks to multiple-beam interference that this striking property has been established.

The high sensitivity only occurs near the maxima, hence it is distinctly advantageous to use a group of perhaps five or six wavelengths instead of a single line such that if t is such as to miss the maximum for one wavelength, it may be nearly right for another.   The high-dispersion contrast technique is therefore even better when used with an unfiltered mercury arc.   One sees uniform areas of different tint.   The eye is particularly sensitive to tint changes, and in the mica case a step of a single lattice spacing makes considerable change in tint.

When this high-dispersion method of enhancing contrast is applied to the study of crystal surfaces a great deal of "invisible" surface structure is brought out into vivid relief.   Both object and matching surface must be correctly silvered, adjusted to be nearly parallel, and also very close together.

Crossed fringes.   Such a high-contrast interference picture cannot be numerically evaluated since there is no evidence available concerning which is the up and which the down (either can give the same intensity change).   Indeed a thickness change dt produces the same effect as one $dt \pm \lambda/2$, $dt \pm 2\lambda/2$, etc., so that what is effectively overlap of orders can occur without there being any easy way of telling this.   One has in fact a most sensitive *qualitative* picture only.   Yet by a simple adaption this can be converted into a quantitative method. The procedure is as follows. The two surfaces being matched (optical flat and object) are set as near as possible parallel to give the high-dispersion contrast-enhancing system. A photograph is taken. Then *without removing the photographic plate* the two surfaces are now slightly inclined to give the familiar sharp wedge fringes, which are then photographed *superposed upon the previous picture*. The result is a "crossed-fringe" pattern of striking power as shown by a typical example in Plate II (c) which is a crossed-fringe system for part of a diamond surface. (Three separate spectrum lines are used in this.) The various twists, turns and steps of the sharp fringes receive an immediate physical interpretation from the contrast picture underlying it. A strong sense of the three-dimensional nature of the topography results. The broad sensitive contrast pattern is now amenable to precision interpretation by the displacements of the sharp superposed fringes. So both systems help each other and a complete evaluation of the topography becomes possible. It is emphasized that the background pattern in Plate II (c) is an interference fringe pattern.

Perfection of silver contour. It has been established by many different lines of approach that a silver layer deposited on a surface by thermal evaporation in a good vacuum has remarkably perfect contouring properties. There is no doubt at all that such a silver film faithfully contours any existing simple or complex surface microtopography. Although the silver film is 500 Å thick, it retains on its surface structural depth details down to perhaps 10 Å if not less. One knows how a thick carpet of snow on a shallow step retains the character of the step on its upper surface. So in the case of evaporated silver. Indeed there is some experimental indication that the first few atomic layers of silver which deposit on a clean surface may be somewhat mobile. It is as if a two-dimensional gas of silver atoms ran freely over every nook and cranny and contoured exactly. Later layers "fix" this and then the final film builds up. This film is but a few hundred atom-layers thick when completed.

Multiple-beam reflection fringes. Up to now we have considered largely only the multiple-beam systems in transmission, which implies that the object under study is transparent. It is evident that *reflection* fringes are the more comprehensive in that they can be used both for opaque and for transparent substances. An extensive application to opaque substances has been made and this includes opaque crystals, metals and a wide variety of other objects, an extreme example, for instance, is polished coal.

In reflection interferometry, the object under study can be coated with a silver film too thick for transmission (to give maximum reflectivity) since it is the lower component of the interference system.   Resting on this is the matching flat which is silvered and through which light penetrates to the object and then returns for observation.   It has already been shown that the *visibility* of the reflection fringes is critically determined by the absorption of this top silvered film, and considerable care in preparation is required, much more so than in the case of transmission fringes.   That adequate definition and contrast can be secured is shown by Plate I (b).

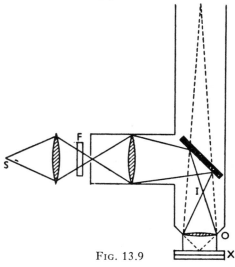

FIG. 13.9

For low-power studies there is no difficulty in observation, but when an object is under study with higher microscope powers a special optical difficulty arises.   It is necessary to use a surface-illuminating type metallurgical microscope.   The object-lens for higher powers is too close to the interference system to permit introduction of a 45° glass reflector between objective and object. The light system used must be as in Fig. 13.9.   Here a mercury-arc source S is filtered by F, passes into the microscope and there meets a sheet of glass at 45° which sends light down to the object X via the objective O.

Now it is essential to use incident parallel light and this is secured by arranging to have an image of the point source formed at I, which is the back focus of the objective O. Thus O now gives a parallel beam on X. At the same time O must be distant from X at the correct microscope image forming distance to give a focused image in the eyepiece. The demands on the design of O are thus severe, for simultaneously it must form a good image from the one side and give a parallel beam from the other. In a typical lens the corrections are not compatible for these two demands and only certain lenses can be used. However, the problem has been solved, and high-magnification fringes can be secured in reflection and have proved useful in technological problems.

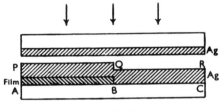

FIG. 13.10

Measurement of thickness of thin films. Reflection multiple-beam fringes have afforded a direct accurate method for measuring the thickness of thin metal films. Such films, varying say from 50 to 5000 Å, are used in a wide variety of physical experiments both electrical as well as optical. Earlier methods for measuring these were: (a) elaborate polarization procedures; (b) weighing, when possible; (c) chemical mass estimation. Not only are these methods difficult but they involved the assumption that the density of the film is that of the bulk metal. The interferometric method gives a direct metrical thickness and is carried out as follows. The film AB (Fig. 13.10) whose thickness is required is deposited on a flat surface, to cover only part of this surface. Over this is deposited, by evaporation, a film of silver, some 700 Å thick. It has been well established that such deposition contours the underlying surface accurately. There is now a step in height separating PQ and QR which is the same height as that of the thin film AB. This composite surface is now matched, slightly tilted, against a silvered optical flat. This is illuminated with parallel monochromatic light, from above so that multiple-beam reflection wedge-fringes are secured. The step appears in the fringes and can be measured with high accuracy. Plate II (d) shows the appearance of the displaced fringes for a step of 550 Å. Film thickness of as small as 20 Å can be measured this way, for the fringes are extremely narrow.

The *reflection* fringes are used because, if light be transmitted through the two separated regions PQ and QR to give transmission fringes, PQ being composite and of different thickness to QR would introduce additional phase change effects by virtue of the differing thickness and this would produce uncertainty. Reflection fringes completely avoid this.

CHAPTER 14

MULTIPLE-BEAM INTERFEROMETRY
II. FURTHER DEVELOPMENTS

Fringes of Equal Chromatic Order

The previous chapter has been concerned with multiple-beam fringes produced by monochromatic radiation. It will now be shown that there are, in certain cases, appreciable advantages to be gained by the use of a different type of multiple-beam fringes which employ white light. These were described by Tolansky (1945) and were called fringes of equal chromatic order. The precision and variety in range and application often surpasses that shown by other types of low-order interference.

The optical set-up is simple, and despite the fact that no monochromatic light is used the fringes obtainable are very sharp, in that fringe width occupies a very small fraction of an order. It can be demonstrated that the fringes of equal chromatic order can be obtained inherently slightly sharper than wedge-fringes, particularly at higher microscope magnifications.

The fringe shape in multiple-beam interferometry is controlled by the Airy expression:

$$I = \frac{I_{max}}{1 + F \sin^2 \delta/2}$$

in which δ is $2\pi(2\mu t \cos \phi/\lambda)$.

For air films μ can be taken as unity. If phase conditions are such that the Airy summation can be approximated to, it is clear that δ is determined by the variables t, ϕ, λ. Combinations of these can lead to the four types of fringes shown below:

Nature of Light	Constant quality	Fringe Type	Name	Filter action
Monochromatic i.e. λ constant	t	Equal inclination	Fabry-Perot	Angular
	ϕ	Equal thickness	Multiple-beam Fizeau	Linear
White-light, i.e. λ variable	ϕ	Equal t/λ	Equal chromatic order	Wavelength
	t	Equal $(t \cos \phi)/\lambda$	White-light Fabry-Perot	,,

Consider first monochromatic light, i.e. λ is constant, leaving t or ϕ as the possible variables. When t is constant (i.e. surfaces are parallel) *fringes of equal inclination* result from variation of ϕ, i.e. from beams of different incidence, Fabry-Perot rings being typical examples.

This optical system behaves in effect as an "angular filter" permitting light to pass which is incident at those angles which correspond to successive orders, n integer.

When ϕ is constant (preferably normal incidence) and t varies (i.e. a wedge), we get *fringes of equal thickness*, multiple-beam Fizeau fringes being typical. Such an arrangement acts as a "linear filter", passing light along successive line regions only.

When white light is used, then λ enters as a variable. Fringes can now be seen when the white light is dispersed with a spectrograph. Taking first the case where ϕ is constant (i.e. preferably normal incidence), then, if t and λ both vary, fringes will result when t/λ is constant for each fringe, since $2t/\lambda$ now represents the order of interference and each order defines a fringe. It will be recognized that any fixed value of t will pass a number of discrete wavelengths, for each of which n is integral. The Edser-Butler fringes described earlier are a special case which arises only when both t and ϕ are constant and λ is the only variable. In the *fringes of equal chromatic* order both t and λ may vary independently.

When t is constant and λ and ϕ vary we can have *white-light Fabry-Perot fringes* if an image of the Fabry-Perot interferometer be projected with a lens on to a spectroscopic slit. These will not be discussed here. They consist of equally spaced curved fringes, and are best observed with a projecting lens of quite short focal length, e.g. a 1-in. microscope objective.

Optical Arrangement for Fringes of Equal Chromatic Order

As in the case of Fizeau fringes, the same collimation limitations and the same restrictions as to phase lag apply. However, as will soon be apparent, it is possible to reduce phase lag to be very small in many cases. The optical set-up is that shown in Fig. 14.1.

A is a white-light source, an image of which is projected by a lens B on to a circular aperture C (some 2 mm in diameter, the value not being critical). C is at the focus of an achromat lens D, which projects a parallel beam of white light on to the interference system E, the incidence being normal (if non-normal incidence is used the phenomena are complicated by the differential polarization phase-change doubling).

With the good achromat lens F, or a microscope objective, an image of the film E is projected on to G, the slit of a prism spectrograph. The fringes appear at H, where they can be observed with an eyepiece or photographed. A reproduction of the fringes is shown in Plate III (a) which covers part of the visible region from the red to the green and has a wavelength scale and a matching spectrum superposed. These fringes correspond to the narrow section of the interference film E, which has been selected by the slit G, since an image of E is projected on to G.

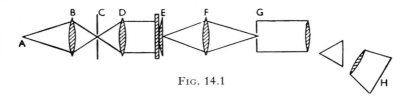

FIG. 14.1

Attention may be drawn to the following special features:

(1) The separation between fringes, in ångström units, is considerable.

(2) The fringes are sharp and narrow, occupying but a small fraction of the wavelength interval separating a fringe pair.

(3) Each fringe exhibits a fine structure.

(4) Each fringe is similar to its neighbours.

(5) Each fringe is multicoloured and can extend over some thousands of angströms, according to the magnification of the lens F.

For each of the curved fringes $2t/\lambda$ is constant. This is the order of interference for this particular fringe, yet each fringe is chromatic; indeed it is possible to obtain a single fringe which sweeps in a curve right through from the red to the blue and back again. For this reason the name fringes of equal chromatic order has been adopted. Each point on any one fringe corresponds to a different path length (t) in the film, yet each contains the same number of waves for the wavelength varies regularly from point to point to compensate for variation of t. If t varies in a complex manner, the fringe shape follows it faithfully. The spectroscope

PLATE III

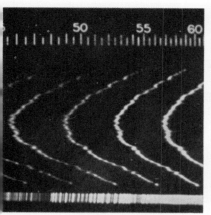

(a) The fringes of equal chromatic order given by a silvered Newton's rings lens-plate combination.

(b) Fringes of equal chromatic order given by a doubly silvered thin sheet of mica. Each fringe is double, due to crystalline birefringence, the two components being plane polarized, mutually perpendicularly. Cleavage steps are shown.

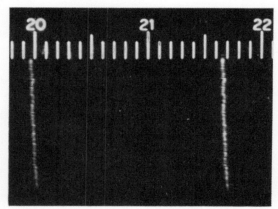

(c) Fringes of equal chromatic order given by a small region of a pair of polished optical flats.

(d) Fringes of equal chromatic order for a small depression on a surface, depth some 4×10^{-6} cm.

PLATE IV

(a) Multiple-beam localized fringes given by a rectangular quartz plate maintaining oscillations. Amplitude of oscillation exceeds a fringe order, leading to overlapping.

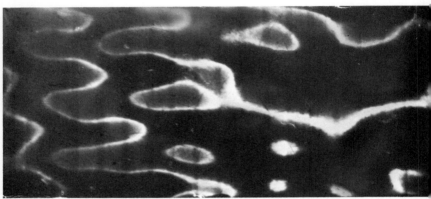

(b) Stroboscopic fringe picture of an oscillating quartz plate. The illuminating source oscillates at the same frequency as the plate.

(c) Multiple-beam localized fringes given by an oscillating quartz disc. Low dispersion is used to increase the number of fringes. The pattern resembles Chladni figures, but in addition, the fringe broadening permits the local amplitude to be evaluated.

(d) Localized multiple-beam fringes given by an oscillating quartz "clock" in the form of a ring annulus. Local nodes and local amplitude variations are revealed.

slit is in effect selecting a narrow section of the interference system and each fringe is a contour of the variation of t in that section, for there is exact point-to-point correspondence between the selected region and its image on the slit.

Dispersion

Since $n = 2\nu t$ the wave-number separation between orders at any horizontal section is $\Delta \nu = 1/2t$. Thus it is only necessary to measure $\Delta \nu$ at any section to secure the value of t at that region on the interference system. As a result of this simple relation there is a very important distinction between fringes of equal chromatic order and Fizeau fringes. In the latter, fringe dispersion is determined only by the *wedge angle*. If high magnification is to be used (on a small region of an object) it is *necessary with Fizeau fringes to increase the wedge angle considerably* in order to bring down the dispersion sufficiently small to accommodate a few fringes in the very restricted field of view of the microscope. This increase in angle seriously increases the phase-lag effects, the displacement of beams and the local value of t, hence definition suffers accordingly.

Now on the contrary the separation between fringes of equal chromatic order has nothing to do with the wedge angle, but as $\Delta \nu = 1/2t$ it is *fixed by the absolute value of t only*. Indeed, even for the highest microscope powers the wedge angle can be zero. The two surfaces can be placed as parallel as possible, so parallel that only the inherent surface topographical angles themselves operate. Since this angle can be kept as low as possible, phase effects are reducible to negligible proportions, with the result that high definition is obtained.

To increase dispersion it is only necessary to bring the interfering surfaces as close together as is possible. With a parallel-sided interference system, or with a wedge whose intersecting edge is parallel to the slit, alteration in the magnification of the lens system F (which can be a microscope) makes no difference at all to the positions of the fringes, which are determined solely by the local value of t for the region imaged on the slit.

It will be noted that the fringes show local irregularities. These are due to the defects on the glass surfaces. As $n\lambda = 2t$ and n is constant for a fringe, then $\mathrm{d}t = n\mathrm{d}\lambda/2 = t\mathrm{d}\lambda/\lambda$.

Thus small changes in t can be derived from measuring the small changes in λ. When the dispersion is not too large and displacements are small it is often simpler to measure dn as a fraction of an order displacement taking a mean value for λ as that of the fringe itself when $dt = dn \cdot \lambda/2$.

Direction of Displacement

One great advantage of these white-light fringes is that no ambiguity can arise as to whether one is concerned with a hill or a valley. Thus circular Newton's rings appear equally from a convex lens or a concave lens and allocation of the correct topography is then not possible merely from a single photograph. With fringes of equal chromatic order one would in both cases see parabolic fringes. But in the convex lens the value of t at the lens pole is less than elsewhere, hence the fringe separation is greatest at the centre and the fringes are *convex* to the violet. With a concave lens, the reverse is true, the fringes pack together at the centre and are *concave* to the violet. Thus at a glance one distinguishes between hill and valley regions.

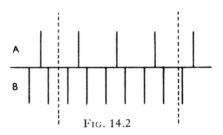

FIG. 14.2

Still more important is the ability to give an unambiguous answer at discontinuities. A sharp step-like discontinuity with monochromatic fringes cannot be evaluated at all, since one neither knows which order belongs to which, nor can one tell which region is up and which down. With the fringes of equal chromatic order the answer is given thus. At a step region the higher part, i.e. that nearest the matching flat, is involved in a smaller t than the lower. The fringes will therefore appear as in Fig. 14.2. For the higher region the dispersion is greater than for the lower, and in this A is the higher and B the lower region.

The system can be evaluated thus. Let any two selected wavelengths Δv apart be chosen (dotted lines). Let there be m orders on the one side of the dividing line and n on the other. (*m* and *n* are in general not integers and include fractions of orders.) Let the distances to the optical flat for the two separate regions be t_1 and t_2, then

$$t_1 = m/2\Delta\nu$$

$$t_2 = n/2\Delta\nu$$

The step $t_2 - t_1 = (n-m)/2\Delta\nu$. Thus both direction and value of the step are determined. Plate III (b) illustrates a typical cleavage discontinuity to which this calculation is applicable.

Measurement of Birefringence

An interesting application of these fringes has been made to thin sheets of mica silvered on both sides. Here for the first time we are considering another variable μ. Now mica is a birefringent crystal and as a result, when fringes of equal chromatic order are secured, it is found that each fringe is double, the two components being plane polarized mutually perpendicularly. The doubling arises from the fact that μ is different for the two directions of vibration.

Since $$n\lambda = 2\mu t$$

then $$nd\lambda = 2d\mu t$$

i.e. $$d\lambda = d\mu.\lambda/\mu$$

Thus the birefringent wavelength doubling $d\lambda$ is *entirely independent of t*. This is a most unusual feature in that we have an interference effect arising in plane parallel plates which is independent of plate thickness.

The value of $d\lambda$ remains the same for thick and thin specimens. The wavelength separation between orders $\Delta\lambda$ depends on the thickness hence the fraction of an order $(d\lambda/\Delta\lambda)$ occupied by a birefringence doublet increases with thickness.

Plate III (b) is a typical example of the doubled fringes given by a doubly silvered thin sheet of mica.

The fringes of equal chromatic order given by a small region of a pair of optically polished flats are shown in Plate III (c). Superposed is an arbitrary scale in the spectrograph which assists in wavelength identification. The fringes represent the detailed structure over a region 1 mm long on the flats, the magnification (on the original) being some $\times 50$. On a typical such interferogram the fringes may be 100 mm apart and fringe irregularities of less than 1 mm in extent can readily be seen and measured.

However, a change of fringe order by one unit corresponds to a height change on the surface of some 2500 Å, and as one-hundredth of this, at least, is measurable it follows that surface irregularities as small as 25 Å can be revealed and measured. It will be noticed incidentally that these particular surfaces are surprisingly uniform over such small regions as 1 sq. mm. The surfaces can thus be used with confidence for matching against others for the purpose of studying microtopography.

Plate III (d) is an example of high-definition fringes of equal chromatic order for a small surface feature, a hump some 4×10^{-6} cm high. This picture illustrates the remarkable magnifications which the interference system renders available. For these fringes, on the original photograph, are some 15 cm apart. Yet as the real separation on the surface corresponding to the two orders is some $2 \cdot 5 \times 10^{-5}$ cm (i.e. half a light wave) the magnification in the fringes is no less than $\times 600\,000$! This enormous magnification is by no means empty magnification for it is accompanied by great resolution, but of course only in the up-down direction.

Interferometric Studies on Crystal Oscillations

The application of the piezo-electric effect to the production of oscillations in crystals, particularly quartz crystals, plays a most important role in many electro-technical procedures. Oscillating circuits can have their frequencies precisely controlled by incorporation of an oscillating quartz crystal into the circuit. High-frequency radio and most radar equipment uses crystal frequency control. So stable is the oscillation of quartz that such a crystal can be used as a clock with a precision exceeding that of the best and most costly of mechanical clocks. Indeed from a group of such "quartz clocks" slight irregularities in the spin of the earth on its axis (the length of the day) have been detected. Then, again, piezo-electric oscillators are used at relatively low frequencies to produce ultra-sonic sound waves, which constitute a formidable field of application and of study.

The frequency at which a quartz crystal oscillates depends upon its dimensions. A thin small crystal can easily be induced to oscillate at a megacycle (10^6 cycles) per second or more. As a rule the amplitude of oscillation is relatively small, and calculations of amplitude can be made from known elastic properties and the applied potential.

In view of the importance of such oscillators it is of some

interest that interference methods can be applied to the study of such vibrations. Results of considerable interest were first reported by Dye (1932) who used two-beam interference methods. Far more accurate and precise information can be revealed by the use of multiple-beam methods.

It is somewhat surprising that the highly sensitive multiple-beam fringes can be applied directly to surfaces making rapid oscillations. The success of this experimental approach is due to the extreme stability of crystal oscillations which can maintain a given distribution of nodes and antinodes with astonishing fidelity over long periods of time. The use of multiple-beam wedge-fringes for this purpose was first used for oscillating crystals by Tolansky and Bardsley (1948), whose techniques will now be described.

The quartz crystal, in the form of a thin slab, has its two major surfaces optically polished to be flat to within a fraction of a light wave. One surface is silvered and this rests lightly on a larger silvered optical flat. Electric contact is made to the silver on the flat. Above the crystal and slightly away from it is a wide mesh of thin wires. The silver on the flat and the wire mesh are the two electrodes to which the alternating potential is applied which sets the crystal into oscillation. By varying the tuning a crystal can be induced to vibrate in any one of several distinct modes.

Multiple-beam wedge-fringes (green mercury light) are produced between the silvered face of the crystal and the flat when the latter is at rest. The crystal is then set into oscillation and a beautiful fringe pattern results, which is quite stable and permits itself to be photographed over minutes, or even hours. Some parts of the crystal oscillate and some are at rest (nodal points). Where the crystal is oscillating the crystal-flat separation increases and decreases, hence the fringes broaden locally. The local fringe width is an exact measure of the local amplitude of oscillation. A width extending over an order means $\lambda/2$ amplitude of oscillation. A typical example of the interferograms given by an oscillating plate is shown in Plate IV (a). Dispersion is arranged to be high, to give only a few fringes over the plate. Vibration amplitudes exceed one interference order such that fringes run into each other.

It is important to note that such an interference picture *only reveals the component of oscillation perpendicular to the surface.*

The amplitude may be perhaps of the order of half a light wave. Clearly if there are movements of a similar nature *along* the surface (and there often are, especially when crystals have oscillations due to shearing forces) they will not be revealed even by high-power microscopy. The regular distribution of nodes and anti-nodes is strikingly demonstrated.

The interferogram only reveals the nodes, antinodes and the amplitudes. It gives no information as to *phase*, i.e. as to which region is going up when another region is going down. This problem has been solved by a stroboscopic technique. The light source used to produce the fringes is a high-frequency electrode-less discharge in mercury vapour. A fraction of the oscillating voltage used to drive the crystal is tapped off and excites the source. Crystal and source are now oscillating at the same fre-quency, with the result that a "frozen-in" stroboscopic picture of the distribution of oscillation is revealed by the interference fringe pattern. It is an easy matter to introduce a controlled phase difference between light-source volts and crystal-oscillation volts, and thus the stroboscopic picture can be moved at will to various sections of the oscillating pattern. Plate IV (b) shows a typical interference picture taken by this stroboscopic technique. The stroboscopic action is not quite perfect and some fringe broadening results in places.

An interesting variation of the direct interferometric method involves deliberate use of low dispersion. Wedge angles are made large such that many close-packed fringes cross the field of view. When oscillations are set up the local broadening at anti-nodes leads to a pattern somewhat resembling the familiar sand patterns given by the Chladni plate experiment in acoustics. Indeed the fringes are essentially revealing the Chladni diagram of the oscillating quartz crystals, with the added information as to numerical values of amplitudes. Plate IV (c) is a typical example of the pattern revealed by an oscillating quartz disc.

Finally, Tolansky and Bardsley, have succeeded in securing the interference pattern given by an oscillating quartz clock. For reasons of stability it has been found that the best and most reliable quartz clocks consist of a thick annulus of quartz cut from a disc, with a smaller inner disc removed. This has to be sup-ported at three nodal points. To secure a controlled dispersion multiple-beam interference pattern over so awkwardly shaped an object is by no means easy. Plate IV (d) shows such an oscillation interferogram given by a quartz clock crystal belonging to the National Physical Laboratory, Teddington, England. The number of nodes, the symmetry of their distribution, and the local variations in amplitude are all revealed in a beautiful manner.

These experiments show that multiple-beam interference methods can offer much in connection with this important field of crystal physics. For instance the extreme sharpness of the nodes (comparable to the fringe width) proves that the nodal regions are at rest to within less than 50 Å during the whole time of observation. It is this unique stability which permits so sensitive a technique as multiple-beam interference to be used on a rapidly oscillating body.

It is of incidental interest to note that the *velocities* involved in these oscillations are very small, even although the frequencies are high. The period may be 10^{-5} sec but the amplitude may be only 10^{-5} in., i.e. a velocity of the order of merely one inch per second!

Interference Filters

A Fabry-Perot interferometer with very small t is effectively a wavelength filter. When such an interferometer is placed before a spectrograph and illuminated at normal incidence with a parallel beam of white light, fringes are passed at different wavelengths, the separation between orders $\Delta \nu$ in wave numbers, being $1/2t$. As already indicated the Edser-Butler fringes make use of this. As t is diminished the fringes separate farther and farther apart. Since $\nu \lambda = 1$ the wavelength separation between orders $\Delta \lambda$ is $\lambda^2/2t$. If the interferometer consists of silver films enclosing refracting material of index μ, then t must be replaced by μt. (If the material has strong dispersion, i.e. μ varies a good deal with λ, complex corrections require to be made.)

Now as t is diminished the fringe separation can be so increased that it is evident that in due course a stage will be reached where there is *one* fringe in the visible region, the next on the shorter wavelength side being in the ultra-violet with the next on the longer wavelength side being in the infra-red. In other words *the system acts like a wavelength filter* effectively passing only that region covered by the width of the fringe, for very little light passes between the orders.

For this to happen t has to be very small indeed. Suppose the interferometer encloses material of refractive index $\mu = 1\cdot5$.

Then t must approximate to the dimensions of half a light wave. Thus take $t = 3 \times 10^{-5}$ cm, then $n\lambda = 2\mu t$, i.e. $n\lambda = 9 \times 10^{-5}$ cm.

On substituting successively $n = 1$, 2, 3, these first three fringes appear at values of λ which are respectively 9000 Å, 4500 Å and 3000 Å. This is precisely the required condition, for such an interferometer is passing only one fringe in the visible region, i.e. it is a "monochromatic" filter (using "monochromatic" in a broad sense) for region 4500 Å. By suitably altering t the position of this fringe in the wavelength scale can be moved. The degree of monochromatization is fixed by the fringe width, i.e. by the reflectivity.

Now if very high reflectivities are used the fringe half-width can be about one-fiftieth of an order. The fringe-order separation in wave numbers is 11 000 cm^{-1} in this special case. One-fiftieth of this is 220 cm^{-1}. This represents the band width transmitted. At 4500 Å this equals 45 Å. The system is therefore a "monochromator" in the sense that it passes a band of half-width only 45 Å even when illuminated by white light.

Strictly speaking, since the Fabry-Perot minima never become quite zero, there is *some* light passed over the whole spectrum, but the amount between orders is very slight provided the reflectivity is high. The higher the reflectivity the narrower is the band-pass, but at the same time the poorer is the intensity transmitted, since the absorption of the silver films increases rapidly when the reflectivity rises.

Fabry-Perot interference filter monochromators were developed in 1942 independently by Tolansky in England and Geffken in Germany, neither being aware of what the other had done because of war conditions. The earlier filters were made as follows. On a glass plate, *which need not be flat*, silver is deposited by thermal evaporation. Then, without breaking the vacuum, a thin carefully controlled layer of cryolite (sodium aluminium fluoride), or magnesium fluoride, is deposited by thermal evaporation on to the silver. Finally a second silver layer is put down, and the thin "sandwich" of silver-cryolite-silver constitutes the filter. So perfect is the contouring mechanism, that a highly uniform filter forms even on a glass surface irregular in the interferometric sense.

The All-dielectric Filter

The earlier filters, if of narrow band-pass, had relatively poor transmission because of the absorption by the silver. A considerable improvement in transmission has resulted from the replacement of the silver by complex multilayers of alternate high-low refractive index, in the manner already described in a former chapter. Beginning with glass, there is deposited on this as many as up to nine or more layers, alternately of cryolite and zinc sulphide. This multilayer is the high reflector. Then the "spacer" (of cryolite) is put down and the thickness of this determines the wavelength passed. This spacer is followed again by a nine-layer multilayer reflector. Such a filter consisting of nineteen layers has been reported to have a half-width as low as some 20 Å, yet it can transmit 70% of the incident light (of that wavelength which it passes).

Frustrated Total-reflection Filter

An ingenious type of interference filter, with transmission exceeding 90% was described by Leurgans and Turner (1947). In this a thin layer of magnesium fluoride is deposited by evaporation on to the hypotenuse of a right-angled, 60°, 30°, glass prism, made of glass of high refractive index. This is followed by a zinc sulphide spacer, which is again followed by magnesium fluoride. To the outer fluoride film is cemented another right-angled, 60°, 30°, glass prism, with a cement having the same refractive index as the glass. The arrangement is shown in Fig. 14.3. Light entering the block from above meets the hypotenuse at an angle exceeding the critical angle for total internal reflection. If, indeed, the magnesium fluoride film were thick compared with the wavelength of light, there would be *complete* total internal reflection. But it is well known from electromagnetic theory of light that at total internal reflection there is *some* surface penetration. The magnesium fluoride is arranged to be so thin that appreciable light passes through it. The total reflection is no longer complete but some *frustration* occurs. By suitably modifying the film thickness, any desired percentage reflection and transmission can be secured, *and there is no loss by absorption* since the very thin fluoride layer is effectively 100% transparent.

The thickness of the zinc sulphide spacer determines the location of the band-pass. Effectively the system is that of a thin spacer between two extremely efficient high reflectors. The system actually transmits two separate wave bands, each plane polarized, mutually perpendicularly.

Interference monochromators are now finding more and more

application.   By attention to experimental detail they can be made for any desired wavelength, a great advantage over other types of conventional colour filters.   One recent example of their use illustrates their versatility.   Goggles were constructed to pass

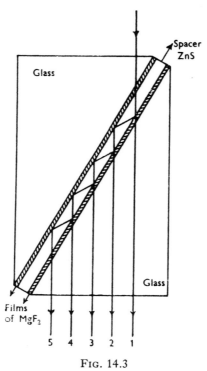

FIG. 14.3

the wavelength of the green auroral line.   When the northern night sky was viewed with these during an auroral display, an accurate picture of the auroral distribution was revealed and could be sketched.

Fringes Formed by Curved Thin Sheets

The interferometric systems described up to now have involved either plane surfaces, approximate planes, or curvatures usually of considerable radius.   In this section a brief survey will be given of the formation of very sharp localized multiple-beam fringes, produced by bending a doubly silvered thin sheet into a complete semi-cylinder.   These fringes were first described by Tolansky

and Barakat (1948) who used as a convenient material a thin sheet
of mica, silvered on both sides and bent into a cylinder of radius
2 cm.   In their simplest form the fringes are produced by monochromatic light, according to the scheme in Fig. 14.4.

A parallel beam of monochromatic light is incident on the convex face of the mica.   Related but different effects result if mica
is turned so that the light falls on the concave face.   A suitable
mica thickness, t, is some 0·02 mm and a suitable cylindrical
curvature has a radius, R, of perhaps 2 cm.   Fringes can also be
obtained over relatively wide ranges of both t and R.   The origin
of formation of the fringes is shown by Fig. 14.4 where for

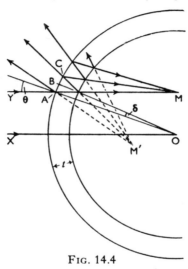

FIG. 14.4

simplicity refraction effects within the interference film have been
disregarded.   To a close approximation both faces can be considered to have the same radius of curvature.   Parallel incident
light at any specific point A is incident on the film at an angle θ
which is determined by the distance XY above the axis of symmetry XO.   Provided t is small compared with R (in the case
quoted the ratio is $1:1000$) then for any particular incident ray,
YA, the successive multiple reflections meet at a point M and there
interfere.   If θ, R and t are such that the optical path difference
between successive beams (including phase-change effects) is
an integral number of waves $n\lambda$, then reinforcement occurs at M
and a bright fringe results.   Clearly at a farther distance from

XO there is another ray with a suitable angle of incidence leading
to another fringe with path difference $(n-1)\lambda$. Hence a succession of localized fringes form and owing to the multiple-beam
effect these are very sharp.

Fringe Localization

An approximate geometrical analysis serves to prove that when t
is small the fringes are localized on a plane passing through the
centre of curvature O and perpendicular to the axis of symmetry
XO.

An example of the fringes is shown in Plate V (a), which is an
illustration of one half of the field of view.

Attention may be drawn to the following features: (i) the
fringes are sharply localized in the predicted plane; (ii) they are
very narrow, the outer orders becoming progressively sharper;
(iii) successive orders clearly follow some simple geometrical law
of separation; (iv) the fringes are double and the doubling exhibits
an unexpected variation with order. These features will now be
discussed in turn.

A striking characteristic is the unusual sharpness of the fringes
for a given silvering. The reason lies in the increased reflectivity
resulting from the large values of the angles of incidence θ which
vary over the whole range from $0°$ to $90°$. It is this increase in θ
which is directly responsible for the marked sharpening up of the
outer orders by producing high effective reflecting coefficients at
the metallic surfaces. It will suffice to indicate a simple derivation of the fringe separation law for successive orders. The
order of interference n_0, at the centre of the system is a maximum
and is given by $n_0\lambda = 2\mu t$. Consider a symmetrical system with
fringes on either side of the axis XO, then the distance between a
pair of symmetrically situated fringes on either side of this axis
will be called 2ρ, the "diameter" of the fringes. If n is the order
of interference for the pth fringe, counting from the centre outwards, then by analogy from the corresponding Fabry-Perot
theory one obtains the following: Refraction can no longer be
disregarded. Let r be the angle of refraction within the film;
then as $n_0 = 2\mu t/\lambda$, the order of interference at the centre, the
order n for the refracting angle, is $n = n_0 \cos r$; $\sin \theta = \rho/R$, so
that:

$$n/n_0 = (1-\rho^2/\mu^2R^2)^{1/2}$$

In general the order of interference n_0 at the centre will be non-integral; let it therefore be written $n_0 = n_1 + \epsilon$, where ϵ is a fraction and n_1 the order of the first fringe. The order of the pth fringe will thus be $n = n_0 - (p + \epsilon - 1)$.

Hence:

$$\{1 - (p + \epsilon - 1)/n_0\}^2 = 1 - \rho^2/\mu^2 R^2$$

As p is small compared with n_0, then, sufficiently closely,

$$(2/n_0)(p + \epsilon - 1) = \rho^2/\mu^2 R^2$$

giving

$$\rho = R\{(\mu\lambda/t)(p + \epsilon - 1)\}^{1/2}$$

The scale of the fringe pattern is therefore directly proportional to R and inversely proportional to $t^{1/2}$.

This formula has been confirmed by measurement and shown to hold closely. By assuming for simplicity that n_0 is integral (i.e. $\epsilon = 0$) the scale of the pattern can be readily calculated. Thus, for example, for $R = 5$ cm and $t = 0 \cdot 02$ mm the diameter of the fifth fringe ($\lambda = 5460$ Å) is almost exactly 4 cm.

The fringes are therefore very easy to see with the naked eye if they are allowed to fall on a ground-glass screen placed in the plane of localization. These fringes are intimately related to the two classical types of multiple-beam interference fringes, namely fringes of equal thickness and fringes of equal inclination; it is desirable to classify them so as to retain this relationship. One way of regarding them is the recognition of the fact that each fringe is formed as the locus of rays which on entering the system make the same angle of inclination with the tangent at the point of entry. These fringes could be called "fringes of equal tangential inclination".

This terminology fits all cases, and if, for example, a spherical shell replaces the cylindrical film (e.g. a blown thin curved film of glass of uniform thickness), then ring-shaped localized fringes arise, since the equal tangential loci then lie on rings.

The relationship to fringes of equal inclination is clear, for the fringes under discussion move off to infinity and become identical with fringes of equal inclination when R becomes infinite, i.e. when the sheet is plane parallel.

When the localized multiple-beam fringes given by an air wedge are formed with non-normal incidence, the fringes split into doublets, the two components being plane polarized mutually perpendicularly. Both become sharper with increasing incidence, but one much more rapidly than the other, and at the same time the sharper fringe weakens rapidly relative to its slightly broader companion.   It has been fully established that this is due to the differential phase change which takes place when mutually perpendicular vibrations of light are reflected at non-normal incidence from a metal, and is in fact the same effect that leads to the elliptic polarization of such reflected light.

Because of the severe bending of the silvered mica sheet, considerable angles of incidence are involved and thus inevitably the outer orders must show the differential phase-change doubling. This is an effect which increases at such a rate that it becomes perceptible at $\theta = 25°$ and increases to about half an order for higher values of θ.

This doubling of the fringes is found even if the interference is produced in an air film bounded by two cylinders.   To it there is now added a further complication if the curved material is a birefringent crystal, as is the case of mica.   For due to the birefrigence the fringes are double, mutually perpendicularly plane polarized.   This fringe doubling diminishes to a small value as the order increases and then begins to increase once more. Over the whole range the outer component becomes progressively sharper and weaker and then effectively vanishes.   Plate V (b) is an example of these fringes given by a mica sheet with several cleavage steps on it.

The observations are completely accounted for in that they represent a combination of birefringent doubling due to the crystal, on which is superposed differential phase-change doubling due to the silver.

In general the splitting of any incident ray by the crystal into two mutually perpendicularly polarized rays travelling with different velocities leads to two systems of interference fringes. Leaving out for the moment the phase-change effect, then the separation between any pair of members belonging to the two systems, but of the same order of interference, is due to the difference in the refractive indices of the two vibrations.   Thus for a biaxial crystal there will be two directions in which the difference in refractive indices is zero.   The actual angular location of the zero positions within the fringe system are decided by the direction of flexure of cylindrical curvature.   Thus the angular separation between the two positions of minimum separation will give double the apparent optic angle of the mica if the differential change of phase at reflection mica-silver is allowed for. Measurements show this to be in fact the case.

The Reflected System

The transmitted fringes have associated with them a reflected complementary system, in which the fringes are dark lines on a bright background. Yet these fringes are unique in that the localization of the reflected fringes differs entirely in character from that of the transmitted system. This will be appreciated at once from Fig. 14.4 in which it is clear that the reflected rays, when traced back, give a virtual fringe image at M'. But M' does not lie on a plane of localization as was the case for the transmitted system. In fact ray tracing shows that the locus of M' is the caustic pattern shown in Fig. 14.5.

If the thin film be curved to be concave to the incident beam, then the localization of transmitted and reflected systems interchange roles when compared with the localizations for the case when the interference film is bent convex towards the incident light.

Fig. 14.5

This fringe system has applications. Local changes in thickness, such as steps due to cleavage, reveal themselves by sharp discontinuities, and accuracy of measurement can be very high indeed. An example is shown in Plate V (b). A thin film can be deposited over part of a mica strip and thus its thickness determined by the fringe shift. The dispersion is variable in that it is determined by the radius of curvature into which the thin film is bent and thus can be altered at will. If white light is used crossed with a spectrograph, fringes analogous to fringes of equal chromatic order can be formed.

Non-localized Multiple-beam Fringes

It has generally been considered in the past that the ring fringes given by the Fabry-Perot interferometer (which normally appear to be at infinity) require an extended source. Strictly speaking this is true, yet as shown by Tolansky (1944), provided certain geometrical conditions are fulfilled, it is possible to produce pseudo-Fabry-Perot rings from a point source. These rings are non-localized, extending out into space and appear on a screen suitably placed. A most striking characteristic of these non-localized fringes is the fact that they can be formed with relatively enormous dispersion, *without the employ of any lenses at all*. Indeed, it is simple to form true interference fringes in space so large that the first ring *can have a diameter of a metre* or even of several metres. This is a rather unique manifestation for optical interferometry.

In the normally accepted use of the Fabry-Perot interferometer the ring system projected by the image-forming lens appears only over the region covered by the image of the source. This source image must therefore be large enough to cover the number of rings required, and this in turn implies that an extended source is to be used. It had apparently not been recognized that a small source (an approximate point source) can lead to the formation of fringes which have an intensity distribution which is effectively that of the Fabry-Perot system. These fringes diverge in cones from the source and extend outwards from the interferometer, into space. If a screen be placed normal to the axis of the system, rings appear on this, and the farther away the screen is taken, the larger become the rings. Rings several metres in diameter can be formed in a large room or in a long corridor.

They require a bright source, since the light-efficiency is low, for no lenses are involved in their production. The green line from a mercury arc is suitable. A simple mathematical treatment will be given later to show how these conical fringes arise, but even without detailed analysis their origin is clear from the following considerations. A property of the Fabry-Perot interferometer is that in effect it behaves as an "angular filter". The positions of the interference maxima are governed by the expression $n\lambda = 2\mu t \cos \phi$. Because of the multiple-beams the fringe maxima are sharp and, in effect, appreciable light is only transmitted by the instrument in angular regions close to those values of ϕ for which n is integral. To a crude approximation it can be considered that only the light emerging along the regions close to these integral values of n is strong, light for other angles of emergence being effectively suppressed. The light, therefore, emerges along cones of a definite thickness depending upon the reflecting coefficient of the interferometer mirrors, provided certain conditions relating to the size of the source and the value of t are fulfilled.

If, now, a screen be placed behind the interferometer, circular

PLATE V

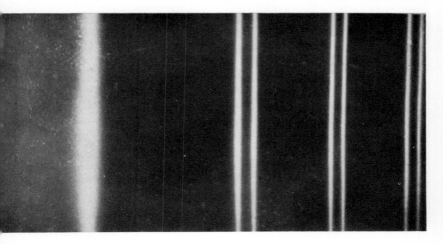

a) Fringes localized on a plane given by a sheet of doubly silvered mica bent into a semi-cylinder. The fringes are double because of mica birefringence, the two components being plane polarized, mutually perpendicularly. Only half the field of view is illustrated.

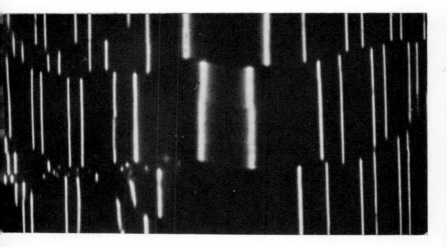

b) Fringes given by cylindrically bent sheet of doubly silvered mica. The sample has cleavage steps on it and this is shown up in the fringe pattern. The rapid variation in doublet separation with increasing order is well shown. The outer component of each doublet is sharper than the inner as its reflecting coefficient is higher.

PLATE VI

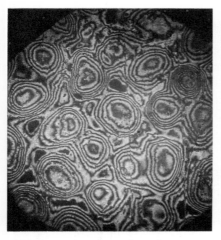

(a) Fringes given by surface of a tomato

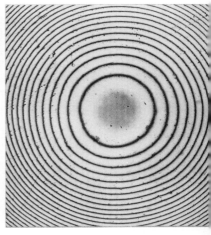

(b) Fringes from steel ball

(c) Holographic interference pattern from belly of
sounding violin

rings appear upon it.   If a certain minimum distance between source and screen is exceeded, the rings formed very closely resemble Fabry-Perot rings.   No lenses are needed for the production of these rings since the conical fringes are non-localized, spreading out indefinitely.   The farther the screen is removed from the source, the larger are the rings.   These rings, formed by the intersection of the non-localized conical fringes with a plane, are not to be confused with the Fabry-Perot rings, which are at infinity and can only be projected on to a screen with the aid of a lens (or seen visually by accommodating the eye on infinity). The appearance of the two types of rings is, however, very similar.

As will be shown shortly, the important practical point for producing these rings is that t should not be large.   It is indeed possible to secure a satisfactory system with $t = 2$ mm, but preferably t should be of the order of 0·1 mm or even less. Convenient spacers for such a purpose are cut pieces of wire or small pieces cut from a sheet of mica.

The Theory of the Ring Formation

The theory of formation of the fringes is as follows.   Let it be assumed that a monochromatic point-source S is situated at some distance from the interferometer PP, the plate separation of which is t (see Fig. 14.6).   As a result of the multiple reflections, the effect upon a screen is equivalent to that produced by a series of equidistant point-sources S, S_1, S_2, S_3 . . . , each $2t$ apart.   If the distance L from the source S to the screen is large compared with $2t$, then to a close approximation the intensities of successive beams reaching any point Q fall off geometrically.

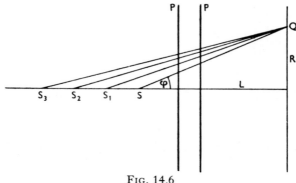

Fig. 14.6

The position of the interferometer relative to the source S is immaterial, provided the aperture of the plates suffices to permit the passage of sufficient beams.

Light from the succession of virtual point-sources leads to the formation of circular rings upon the screen. (A corresponding infinite set of virtual images is formed on the other side of the interferometer. These will lead to the formation of a complementary system of conical non-localized fringes expanding outwards to the left of the diagram. These back-reflected fringes are not considered here.) If the lengths of the paths from the successive points S, S_1, S_2, S_3 . . . were to increase in arithmetical progression, then the rings formed on the screen would have the intensity distribution of Fabry-Perot fringes. Under suitable conditions the deviations of the retardations from this exact condition are not serious.

The path difference at the point Q between the two rays from S and S_1 can be written as:

$$D = \sqrt{(R^2+L^2+4Lt+4t^2)} - \sqrt{(L^2+R^2)}$$
$$= (L^2+R^2)^{1/2}\{(1+\alpha)^{1/2}-1\}$$

where
$$\alpha = \frac{4Lt+4t^2}{L^2+R^2}$$

Expanding the expression containing α to include second-order terms gives:

$$D = (L^2+R^2)^{1/2}(\alpha/2-\alpha^2/8)$$

If SQ makes an angle ϕ with the axis, then powers of t/L greater than the second can be neglected, giving:

$$D = (L/\cos \phi)\{(2t/L)\cos^2 \phi+(2t^2/L^2)(\cos^2 \phi-\cos^4 \phi)\}$$
$$= 2t\phi+2(t^2/L)\cos \phi \sin^2 \phi$$

The path difference D_n between the first and nth beam is obtained if t is replaced by nt, giving:

$$D_n = 2nt \cos \phi+2n^2(t^2/L)\cos \phi \sin^2 \phi$$

For the first ring $D = 2t-\lambda$ making, closely enough, $\sin^2 \phi = \lambda/t$. Thus:

$$D_n = 2nt \cos \phi+2n^2t(\lambda/t)\cos \phi$$

The amount by which the nth ray lags behind the arithmetical progression value is approximately

$$\Delta = (n^2-1)(2t\lambda/L)$$

Suppose $t = 0\cdot1$ mm, and let the screen-source distance L be 1 metre, i.e. $t/L = 10^{-4}$, then the extra path retardation $\varDelta$ between the first and second beams is only 6×10^{-4} of a light wave, an amount of no consequence. Writing $\varDelta = (t/L)2\lambda n^2$ (i.e. replacing (n^2-1) by n^2 for n large), it follows that for this particular t/L, $\varDelta$ has the value $\lambda/2$ for $n = 50$. Thus at a source-screen distance of 1 metre, there is almost complete Airy distribution if t is less than $0\cdot1$ mm.

The above argument, which applies to reinforcement of successive beams with the first beam, can be extended to include reinforcement of any higher-order beam with the series following it.

The Diameters of the Rings

When t/L is small then successive rings 1 to p occur at angles θ_p given by $p\lambda = 2t \cos \theta_p$. A rough approximation as to ring sizes can therefore be obtained from the corresponding Fabry-Perot case in which the same expression holds. The reason why the inference is only approximate is because the small value of t leads to large values of θ and this tends to invalidate the approximation previously used of equating an angle with its sine. However, a qualitatively correct picture will be obtained by adopting as the angular diameter of the pth ring the value $2\sqrt{(p\lambda/t)}$, which would have been the case for the Fabry-Perot interferometer. The linear diameter at a distance L is then $2L\sqrt{(p\lambda/t)}$. Taking $t = 0\cdot1$ mm and evaluating the diameter of the first ring at a distance $L = 10$ metres from the source, for $\lambda = 5000$ Å, leads to the value *1·4 metres*. The diameter of the tenth ring is no less than *4·4 metres*.

It should be noted that these strikingly large interference rings are not large through projection by a lens but the size is an inherent property. (It is clear that this same technique is applicable to radio microwaves, in which case one can have multiple-beam interference microwave fringes with very great diameters easily extending perhaps over 100 metres and more.)

Apart from the question of the approximation involved in replacing an angle by its sine, there is a second-order difference between the sizes of the non-localized rings and those of the corresponding Fabry-Perot interferometer. In effect the value t in the Fabry-Perot case has increased to $t(1+n\lambda/L)$ for the nth beam. Since λ/L is in this case merely 5×10^{-7}, even a value

$n = 50$ merely apparently increases t by 1 part in 40 000. However, the quantity $2t/L$ gives the increase in order of interference for the first two beams for the first ring, as compared with the Fabry-Perot case. In our example $2t/L$ equals $1/50\,000$. Hence the first ring is smaller than the corresponding Fabry-Perot case by a negligible amount.

The fringe width for a point source is practically indistinguishable from that given by the Airy distribution. However, if the source has an appreciable width, then this will lead to broadening; in effect one has to *add source width to line width*. If instead of a single point source two separated point sources are used, then each gives its own separate ring system and these superpose to form a mixed pattern, which is not a composite interference pattern but merely a geometrical intensity superposition. In fact by using slits, squares, triangles, etc., as the illuminated aperture, a whole range of mixed fringe patterns can be produced.

With these fringes the high dispersion is secured with a corresponding high loss in intensity, nevertheless their unique size makes them a fascinating object for demonstration purposes. Intensity can be much improved if a small image of a mercury arc be projected on to the front surface of the interferometer, using a 1-in. microscope objective to form the image.

The fringes have so far found little practical application. They can be produced by silvering a piece of mica on both sides and using it as the interferometer. This then has two especially interesting characteristics. First, because of the birefringence, the fringes emerge doubled, consisting of plane-polarized rings, polarized in mutually perpendicular directions. One set is circular, the other slightly elliptical. This can be explained on crystallographic grounds. Second, if the small image is placed so as to intersect a cleavage line, then as we have effectively side by side two interferometers of slightly different thickness, diverging beams on the one side pass through one mica thickness and on the other side through the other, the two sides differing by the height of the cleavage step. Thus the ring patterns divide at the projected "shadow" of the cleavage line and from the discontinuity the cleavage step can be measured. The usual difficulties occur as to allocation of orders at a discontinuity. The interest in the arrangement lies in the enormous dispersion available with no requirements for auxiliary apparatus. This technique can be applied without ambiguities to the measurement of the thickness of a very thin film. For it is only necessary to deposit the film on a selected uniform sheet of mica, whereby the step height can readily be determined. The sensitivity by this method is the same as that using multiple-beam Fizeau fringes.

CHAPTER 15

INTERFERENCE IN CRYSTALS WITH POLARIZED LIGHT

Introduction

The interference phenomena which can arise in doubly refracting crystals can be highly complex and only the simplest cases will be considered here. When a crystal like calcite exhibits double refraction, a light beam on entering it splits into two beams which are plane polarized mutually perpendicularly. These emerge separated, having travelled in different directions, with different velocities, within the crystal. One beam, obeying Snell's law, is called the ordinary ray and has a refractive index μ_o. The other beam (the extraordinary) has a refractive index μ_e depending on direction. If one imagines a source within the crystal, then the wave-front from this for the ordinary ray is spherical, but for the extraordinary ray it is an ellipsoid of revolution. Two cases arise, one in which the ellipsoid is outside and the other in which it is inside the sphere. The direction which joins the points where these touch is called the optic axis and there is no double refraction in such a direction. If there is one direction in a crystal without double refraction it is called uniaxial. When $\mu_o > \mu_e$ the ellipsoid is outside the sphere and this is described as a negative crystal. The extraordinary ray then makes a greater angle with the normal than the ordinary. When $\mu_e < \mu_o$ the sphere is outside the ellipsoid, a positive crystal results.

Early in the nineteenth century Fresnel and Arago showed experimentally that interference between polarized light beams obeys the following rules:

(1) Two rays of light plane polarized at *right angles* to each other cannot be made to interfere.

(2) Two rays of light plane polarized with planes parallel can be made to interfere.

(3) Two polarized beams brought into the same plane will only interfere if they belong originally to the same beam, i.e. they are coherent, coming from one light source.

198

It follows from this that if a beam of light enters a crystal and is split into two beams polarized in mutually perpendicular planes, these two emerging beams will not interfere unless they are brought into one plane, or resolved components are brought into a common plane, by some polarizing device such as a Nicol prism or a sheet of polaroid. Conversely if they *are* brought into a common plane, then interference will result. We shall consider a few simple cases. One can treat of either a parallel plate of crystalline material or a wedge, and one can consider illumination either by parallel light or by converging light. For each case one can further use either monochromatic light or white light. Some of these cases have practical value.

Interference with Parallel Plates of Crystal

The phenomena depend upon how the plate is cut relative to the crystal axes. Consider a uniaxial crystal, with plane parallel faces, placed between two Nicol prisms, the first of which, the polarizer, causes plane-polarized light to strike the crystal, and the second, the analyser, unites in its plane resolved components of the two rays which emerge from the crystal. The crystal is thin, of thickness d.

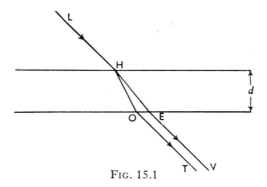

Fig. 15.1

In Fig. 15.1 let LH be the plane-polarized beam striking the crystal. This refracts into HO and HE, the ordinary and extraordinary rays, which leave respectively along OT and EV in what are practically parallel directions. If d is small O and E virtually coincide, and although the two paths HO and HE are *metrically* almost identical and equal to d they are *optically* quite different because of the two different refractive indices. The phase difference between the emerging beams is $(2\pi d/\lambda)(\mu_o - \mu_e)$, λ being the wavelength of the light in air.

The intensity of the two beams HO, HE depends upon the direction of the crystal axes relative to the plane of polarization of the first polarizer. The intensity finally emerging from the analyser again depends upon the direction of the plane of polarization of the analyser. In Fig. 15.2 let OP represent the plane of the initial polarizer, i.e. the plane of polarization of the light incident on the crystal. Let OX, OY be the directions of the planes of polarization of the ordinary and extraordinary rays after emerging from the crystal. Thus OY, OX depend on how the crystal is set relative to OP. If light of unit amplitude strikes the crystal, then the amplitudes along OX and OY are respectively $\cos\alpha$ and $\sin\alpha$.

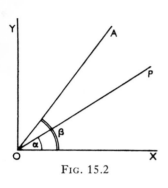

FIG. 15.2

Now let OA be the polarizing plane of the analyser. Only those resolved components of OX and OY along OA will be transmitted by the analyser. These two resolved components have amplitudes $\cos\alpha \cos\beta$ and $\sin\alpha \sin\beta$ and differ in phase by δ. They compound into a single beam whose intensity, as established earlier, is:

$$I = (\cos\alpha \cos\beta)^2 + (\sin\alpha \sin\beta)^2 + 2(\cos\alpha \cos\beta)(\sin\alpha \sin\beta)\cos\delta$$
$$= (\cos\alpha \cos\beta + \sin\alpha \sin\beta)^2 - 2\cos\alpha \sin\alpha \cos\beta \sin\beta(1 - \cos\delta)$$
$$= \cos^2(\alpha - \beta) - \sin 2\alpha . \sin 2\beta . \sin^2 \delta/2$$

When the polarizer and analyser are set parallel, i.e. $\alpha = \beta$ this reduces to:

$$I_{\parallel} = 1 - \sin^2 2\alpha . \sin^2 \delta/2$$

When the polarizer and analyser are at right angles, i.e. $\alpha = \beta \pm \pi/2$, then

$$I_{\perp} = \sin^2 2\alpha . \sin^2 \delta/2$$

The intensities in the two cases are therefore complementary for they add up to unity.

Illumination with Parallel Light

The analyser, as shown above, brings the two beams together into a condition to interfere. If the eye is focused on the crystal plate, then, because small angles are involved, δ is practically the same for all regions. Thus if monochromatic light is used a uniform brightness appears which depends on the particular value of δ, i.e. on the thickness, the difference in refractive indices, and on wavelength.

When white light is used, then as δ is dependent on wavelength, different regions of the spectrum are transmitted differently, according to the thickness, hence a definite pure colour is seen, depending on thickness. It is clear that rotation of the analyser through 90° will produce a complementary colour.

A thin plate of doubly refracting material, such as selenite, can be placed between crossed Nicols and by careful cutting and scraping away of local regions, coloured designs of great beauty can be formed.

The Quartz Wedge

The quartz wedge, of considerable application in polarization microscopy, especially mineralogy, is a thin wedge of quartz, with an angle of about 0·5°. There is of course practically no deviation of light by such a prism. The wedge is cut with the optic axis of the crystal parallel to the sides and in this arrangement the μ_o and μ_e (for sodium light) are 1·5442 and 1·5533, thus $d\mu = 0·0091$.

On illuminating this with parallel light between crossed polarizers, it is clear that δ varies regularly across the wedge, being proportional to the thickness.

Suppose the polarizers are crossed. The intensity formula is then $I = \sin^2 2\alpha . \sin^2 \delta/2$. Let the wedge be set with its axis at 45° to those of the polarizers making $\sin^2 2\alpha = 1$. The transmitted intensity is then given simply by $I = \sin^2 \delta/2$. This is a maximum when $\delta = (2n+1)\pi$, i.e. when

$$d = \frac{(n+\frac{1}{2})\lambda}{\mu_o - \mu_e} = \frac{(n+\frac{1}{2})\lambda}{0·0091}$$

With monochromatic light the wedge is covered effectively with fringes, alternate light and dark bands, whose intensities obey a $\sin^2$ law. When white light is used the different wavelengths have their maxima at different places and the wedge is then crossed by coloured bands in a particular sequence of colours, called the Newton scale, for it is the same sequence as that shown by white light reflected from a thin-film wedge, as described by Newton.

If the analyser and polarizer are turned to be in parallel positions, the monochromatic dark bands change over into light bands and vice versa. Wedges often have a linear scale marked on them to assist localization of fringes.

In a quartz crystal the positive direction is that of the crystal axis hence for the wedge cut parallel with the axis the length of the wedge is the positive direction. The wedge is used to determine the optical characteristics of crystals. In practice the wedge is superposed on the crystal, both set at 45° to the crossed polarizers, white-light illumination being used. If the positive direction in the crystal coincides with that of the wedge, the colour of the quartz will rise in the Newton scale. The crystal is then rotated through 90°, and, as the two now oppose, a position on the wedge can be found which exactly compensates the crystal. The wedge is moved to diminished order of colour until this compensation is found, and then the path difference in the crystal is equal and opposite to that of the region of the wedge being traversed. Thus the path difference for the crystal has been determined. This assists mineralogical identification.

Variations of this wedge interference principle have been developed. The Babinet compensator consists of two quartz wedges, superposed, with crystal axes at right angles to each other, i.e. one along, the other perpendicular to the wedge direction. The zero band is at the centre where the path differences in the two wedges exactly neutralize. On one side of this is negative, on the other positive. There are many other versions.

Interference in Parallel Plates with Converging Light

This effect plays an important part in the study of minerals. Consider a crystal cut at right angles to the optic axis. Let it be illuminated *from below* with a highly converging beam of plane-polarized light OR. The light reaches the eye at E, after passing the analyser A which brings the beams into interference (Fig. 15.3).

For the normal ray OMNE, δ is zero. For any particular semi-cone angle determined by angle NEP, the light reaching E on the surface of this cone has δ constant, for the value of δ is determined solely by the angle of incidence. The eye will therefore see circular interference fringes, which clearly are fringes of equal inclination. They arise because the refractive index μ_e depends on the inclination to the optic axis so that δ increases with increasing inclination.

This simplified arrangement is not used in practice. Two short focal length lenses are employed as in Fig. 15.4. P and A are

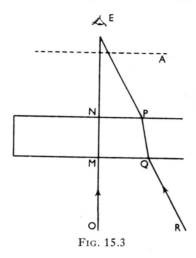

FIG. 15.3

polarizer and analyser. C and O are respectively the condenser lens and objective of a microscope, and the crystal specimen is S. The eyepiece of the microscope is removed and the eye E looks at the back focal plane of the objective O. The fringes are seen there clearly in focus. For fringes of equal inclination appear to

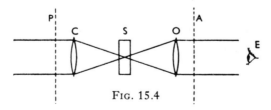

FIG. 15.4

be at infinity and therefore can be seen at the back focal plane of the objective.

When white light is used, coloured rings are seen, since δ depends on λ. With crossed polarizers the intensity is zero for regions where $\sin^2 2\alpha = 0$. This occurs at $\alpha = 0$, $\pi/2$, $3\pi/2$. Thus the circular bright fringes are crossed by a dark cross. The rings themselves are called isochromatic circles (since each has its own colour), whereas the dark crosses are called achromatic lines. If the polarizers are brought into the parallel position, each ring takes on its complementary colour and the achromatic

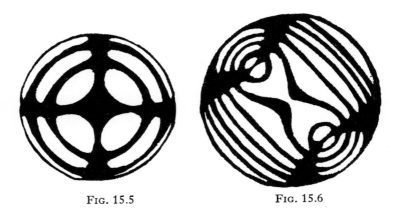

FIG. 15.5   FIG. 15.6

lines become white, i.e. bright. The form of the fringe pattern for a uniaxial crystal is shown in Fig. 15.5, whereas for a biaxial crystal one sees as Fig. 15.6. These interference patterns, some-times called conoscopic figures, are of considerable value in identifying mineralogical specimens.

The Lyot Polarization Interference Filter

A highly ingenious device was first described by Lyot (1933) which permits one, by an interference filter action, to select out from white light a very narrow wavelength band. In the original form such a filter passed a band only 4 Å wide and in later modifi-cations a band of width of only $1\frac{1}{4}$ Å is passed. Such a filter can be of great value especially in astrophysics. Thus the filter wavelength can be arranged to be that of the spectrum line H_α. If the sun be viewed through it, one will see the distribution of solar hydrogen. This permits coronas, solar prominences, sun-spots, etc., to be seen in pure hydrogen light and thus effectively isolates the hydrogen distribution in solar regions. It is impor-tant that the filter should be highly transparent in its own region, since clearly if it only passes say 1/2000 of the visible region it is necessary that there should not be a great deal of absorption in the narrow region transmitted. The Lyot filter achieves this. It is based upon the transmission by a parallel plate of quartz when set between two polarizers, with its optical axis parallel to the face and set at 45° to that of directions of the polarizers. It requires to be made with high accuracy and the few filters made are prized objects.

When considering the wedge (with monochromatic light) one obtained a sin² variation of intensity due to the sin² δ/2 term. Now consider here white light and a plate with thickness constant

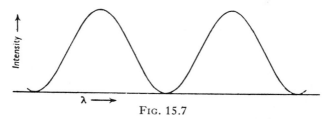

FIG. 15.7

(parallel plate). Again there is variation of δ because δ is a function of λ. Clearly there will be a sin² variation of *transmission* with variation of λ, as represented by Fig. 15.7 which plots intensity against wavelength λ.

Now consider an arrangement as in Fig. 15.8 which consists of four sheets of polarizing material P_1, P_2, P_3, P_4. Between the first and second is a crystal C_1 and then interlaced follow C_2, C_3. Each crystal is cut to be *exactly* twice as thick as its predecessor.

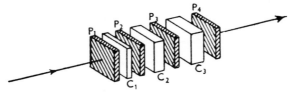

FIG. 15.8

Clearly for a thicker crystal δ varies more rapidly with λ, and in fact the transmission peaks for C_2 are twice as close as for C_1 and for C_3 twice as close again. Thus for each of the three crystals the respective transmissions are as shown in Fig. 15.9 at A, B, C respectively. It is clear that the light which can pass through the whole combination is that shown at D. Hence the system acts as a wavelength filter, selecting the narrow regions in D.

By a crude colour filter one can isolate either of these fringes.

By selecting the correct thickness, the fringe can be localized at some critically selected wavelength.

With six quartz plates ranging from 1·68 to 53·66 mm a band width of 4 Å was obtained for the red hydrogen line λ6563.

There are several experimental difficulties. The thickness must be very accurate. Good-quality quartz of over 5 cm thickness is difficult to find. Temperature control is essential. The system must be immersed in oil to reduce multiple reflections

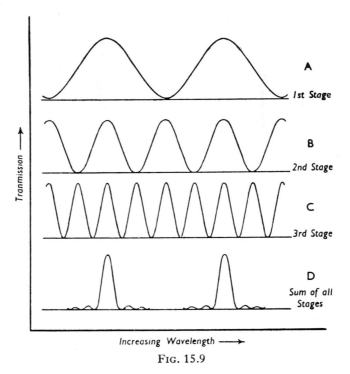

Fig. 15.9

from the various surfaces; even then with all precautions only 10% of the light is transmitted.

More recent modifications (Billings 1947) use ammonium dihydrogen phosphate crystals which have electro-optical properties. Precise temperature control is needed, but a band-pass of 1¼ Å can be achieved. By applying potentials to the crystals it is possible to shift the transmission band as desired, i.e. the filter has a wavelength band-pass under control.

Half-Wave Plate

When a parallel beam of plane polarized monochromatic light falls upon a thin plate of birefringent material, e.g. a section of a suitable crystal, then the state of polarization of the light which leaves the crystal plate is determined by two factors, namely, the path difference between the ordinary and extraordinary rays in the crystal and the angles made by the direction of the plane of polarization of the incident light relative to the polarization directions in the crystal plate itself. The retardation in the plate is proportional to the differences of the two refractive indices and to the thickness.

Let μ_e and μ_o be the two refractive indices for the extraordinary and ordinary rays. When a plate is illuminated at normal incidence with parallel monochromatic light, then, for a crystal plate of thickness t, the retardation between these rays is $(\mu_e - \mu_o)t$. For example, a thin plate of the birefringent crystal mica has $\mu_e - \mu_o = 0.005$ for sodium light. Hence, a mica plate of thickness 0.059 mm has a retardation of $\lambda/2$ for this light and such a plate is then called a half-wave plate. Other crystals, such as quartz, can also be cut into half-wave plates.

When a doubly refracting crystal plate is placed in a beam of plane polarized light the plane of polarization of that light is effectively rotated through an angle by the plate because of the resolved components of the incident vibration into the two vibration directions of the crystal.

It can be shown without difficulty that for a half-wave plate, if the plane of polarization of the incident light makes an angle with one of the vibration directions in the crystal when it enters that crystal, then on emergence it makes an angle $-\alpha$, with that direction. A special case exists for $\alpha = 45°$, for then the plane of polarization of the incident light is on emergence rotated through 90°. Consequently, when a half-wave plate is placed at 45° to the two vibration directions of the extraordinary and the ordinary beams which can emerge from a crystal, then the planes of vibration in these two beams are interchanged. This special property will be exploited later, especially when discussing interference microscopes.

Pleochroism

It has been known for many years that some doubly refracting crystals exhibit a notable difference in their transparency to the two polarized rays, the ordinary and the extraordinary. This is called pleochroism. One long-familiar example is the crystal tourmaline, which is quite absorbing to its own ordinary ray and yet is fairly transparent to its own extraordinary ray. As a consequence of this differential absorption, when unpolarized light is sent through even a fairly thin tourmaline crystal, there emerges only light which is plane polarized (the extraordinary ray) for the light vibrating in the other direction is strongly absorbed.

One man-made crystal which shows very marked pleochroism is Herapathite, a sulphate of iodo-quinine first produced in 1852 by Herepath. This can be made in long thin platelets, and so strong is the pleochroism that even quite thin plates produce, by transmission, almost pure plane-polarized light. This pleochroic property was exploited in a remarkable fashion in 1932 by Land with his invention of polarizing sheets.

Land first impregnated cellulose acetate with microscopic crystals of Herapathite. From the mixture a sheet was extruded and in the extrusion process the thin crystal plates all aligned up with axes parallel. The selected concentration of crystals was such that the plastic sheet almost completely polarized any light transmitted.

Following this material, Land then developed a series of now widely used polarizing sheets called Polaroid H, of which there are now available several varieties. This is manufactured from a plastic sheet of polyvinyl alcohol which is stretched in one direction until its length is increased by 50%. This stretching causes the long polymer molecules which constitute the plastic to orient themselves parallel to each other. The sheet is then immersed for a time in a solution of iodine which reacts with the oriented polyvinyl alcohol molecules, converting them into a strongly pleochroic compound. This sheet of pleochroic molecules now acts as a polarizer. It has numerous advantages over other types of polarizing systems. It is light and strong, can be cut to desired shape and can be made in areas of several square feet, if so required. The cost of manufacture is relatively low and applications are very varied indeed.

The spectral transmissions of Polaroid sheets depend very much on the type selected.

A perfect pleochroic polarizer would have 50% transmission of the one vibration and 0% transmission of the other. In practice, such theoretical values are never attained, but they can be closely approached to in certain wavelength regions. If the transmittance for the main beam which goes through is K_1, and if the transmittance for the nearly absorbed other beam is K_2, then the transmittance of a single polarizer is $\frac{1}{2}(K_1+K_2)$. That of two polarizers crossed (i.e. at right-angles) is K_1K_2 and, of course, since K_2 is normally very low there is almost complete extinction. According to whether maximum transmission or maximum extinction is required, different kinds of material are available. Thus, a single sheet of Polaroid HN-22 transmits something like 22% of the incident light, over the whole visible spectrum (K_1 averages about 0·44), yet the value of K_2 is so low (i.e. 0·000 002) that consequently only a very small amount of light is transmitted by a pair of such crossed Polaroids HN-22. At the other extreme is HN-38 which transmits 38% (instead of the ideal of 50%), but, correspondingly, K_2 is much higher, i.e. 0·005 in the green and is as much as 0·04 in blue. Hence, even when fully crossed, there is some transmission, especially in the blue region.

In addition to these is a type called K sheet which withstands high temperature and actually transmits up to 40%, whilst a useful form, HR, is highly effective in the near infra-red region.

Polaroid sheet is also available which has on one side a thin film of material acting as a half-wave plate and this has certain light-filtering applications. In most optical instruments the older Nicol prism polarizers and analysers have now been replaced by Polaroid sheet. One considerable advantage is that strongly converging or diverging beams can be accepted with little loss in efficiency, which is not the case with the older optical devices.

We shall later describe applications of Polaroid sheet when dealing with interference microscopes.

CHAPTER 16

INTERFERENCE MICROSCOPES

Introduction

A fairly recent development in interference optics has been the improvement and marketing, within recent years, of a number of very good interference microscopes. Such instruments are modifications of microscopes such that whilst an object (possibly either opaque or transparent) is under view the instrument simultaneously behaves as an interferometer. As a consequence of this, additional useful information is made available. It has been possible to adapt high-resolution microscopes for interferometric purposes. The resolving power of a microscope is determined by the size of the collecting cone of light of the objective (the numerical aperture). For high resolution, very short focal length lenses (2, 3 and 4 mm focal lengths) are required. All of these have usually quite small working distances (distance between object and the front of the lens) and this has an influence upon interference microscope design.

For simplicity, interference microscopes can be divided into two groups (i) microscopes for use on opaque objects, such as metals, and (ii) microscopes for use on transparent materials, especially the kind of small objects met with so often in biology. In the main, the instruments in group (i) give information about the structure of the *surface* of an object, i.e. the surface microtopography. In group (ii) the interference microscopes can reveal object shapes, *thickness* variations and refractivity. In particular, they show up structural features with a very considerable enhancement in contrast. The "interference contrast" is a most important technical advance for revealing subtle structural detail, and accounts largely for the increasing use of interference microscopy by biologists. Instruments have become commercially available varying over a considerable range of complexity (and therefore cost). It is characteristic of all interference microscopes that they have high magnification and high resolution *only in the up-down (depth) dimension.* For whilst the optical microscope at its best can only resolve in extension an element of length of some $\lambda/2$, (say at ×1000 magnification) the resolution in *depth* with multiple beam interference microscopes can be of the order of $\lambda/1000$ and magnifications of ×100 000 and more are not uncommon. Thus it is that when objects are examined in the interference microscope, the dimensional scales in extension and in up-down can easily differ by a factor of 100 or even more.

Interference microscopes using reflected light

In general, the microscopes used for examining surface micro-topography either of opaque objects (e.g. metals) or, indeed, also

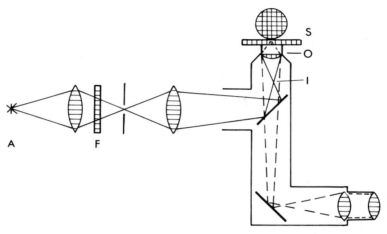

Fig. 16.1

of transparent objects, employ reflected light and are much simpler (and often much cheaper), than the systems needed to view through transparent material (e.g. small biological objects, thin transparent films, etc). The simplest of all the arrangements is that developed by Tolansky (1944) and is shown in Fig. 16.1. It can be used either as a two-beam or as a multiple-beam system. With two-beam interference, good fringe contrast is obtainable with up to a 4-mm objective (say at ×800). With multiple-beams definition falls off when microscope powers exceed a 16-mm lens (say ×200). An inverted type of microscope has advantages, as shown in Fig. 16.1. Light from a mercury arc source A, after being monochromatized with filter F, enters the side of the microscope and a point image is formed at I, at the back focal plane of the microscope objective O.   Because I is at the focus of O the interference system S (here a spherical object is resting on a thin cover slip which acts as the optical flat) is illuminated at normal incidence with parallel light.   The surface of the object is seen in focus and it is covered with interference fringes.   These are, in effect, an optical contour map, height changing by $\lambda/2$ in moving from fringe to fringe.   For many objects two-beams suffice. Thus Plate VI (a) shows the micro interference pattern given with a 4-mm lens ($\times 600$) by the surface of a ripe tomato resting on the optical flat.   The surface consists of a number of cell units which are all hollow depressions.   Both shapes and depths of the depressions are revealed.

For multiple-beam studies, e.g. examination of metals or crystals, then both surfaces involved in interference can be coated by high reflectors.   In certain instances good results follow from the use of high reflecting metal oxides.   Plate VI (b) shows an interferogram for a commercial steel ball-bearing.   Fringes are sharp and contrast is good.   The multiple reflector is a $\lambda/4$ layer of bismuth oxide.

Michelson Interferometer Systems

Several types of interference microscope are modifications of the classical Michelson interferometer.   In 1962 Watson produced an

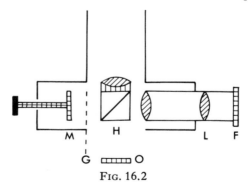

FIG. 16.2

8-mm lens interference microscope objective which is in effect a miniature Michelson interferometer which is built into an 8-mm microscope lens, and this whole arrangement, with adjustments, is small enough to screw directly on to any bench microscope, converting it, at small cost, into an interference microscope, particularly useful for reflected light.

The system is shown schematically in Fig. 16.2. Illumination conveniently is from either a sodium yellow light or a green mercury light source. A magnification of × 300 in extension is possible. Light from the source, filtered by F, is rendered parallel by lens combination L and falls on a beam-splitting cube. The object to be examined (say a specular metallic surface) is at O. The cube, which consists of two right-angle prisms cemented together, with a half reflector H on one hypotenuse, is cemented to the 8-mm microscope objective which has a sufficiently big working distance to permit incorporation of the cube. The mirror M is adjustable for tilt and separation from H. G is a stop by which M can be cut off at will. The Michelson interference system is formed by the reflector H and the two mirrors M and O. When this is set up to give localized fringes then the plane M, together with the microtopographical structure of O, produces an interference pattern which consists of thin film localized fringes which show up numerically the nature of the microtopography. The visibility of the fringes is determined by the intensities of the two beams involved hence by selecting reflectivity of 50% for mirror M this enables the set-up to cope with a fairly wide range of metal reflectivities on O (say 25%–75%) and still thereby obtain reasonably good fringe contrast. Because path differences can be balanced, white-light fringes as well as monochromatic fringes can be obtained.

The beam-splitter, being a cube before the microscope objective lens, automatically restricts the magnification and resolving power available. Several other systems have been designed which have this same defect, but one due to Mirau (1949) is an optical arrangement, closely related in method to one of Michelson's variants, and is one which, in spite of still having the beam splitter between lens and object, yet permits closer approach than in the Watson arrangement. Mirau's device is shown in Fig. 16.3. Using a metallurgical type of incident light microscope, light is brought down by the half reflector R to the microscope objective lens L. This has a small silvered spot at the centre. The beam-splitter S is a half-reflecting film between two identical thin slips of glass. At this divider, half the light goes to the object O and half reflects to the central silver spot. The light returns back to R and through this to an eyepiece. The object is seen covered with interference fringes.   The fact that divider S is normal to the axis and consists of thin plates (not a cube as before) enables closer approach between lens and object than in the Watson device.   A minor drawback is that the silver spot on the lens obscures the lens centre and this affects the ability of the lens to resolve horizontal detail in extension.

In 1933 Linnik described a Michelson-type microscope inter-ferometer, for opaque objects, wherein the beam-splitter is *behind* the microscope objective.   As a consequence of this the *highest* microscope powers, even up to and including oil immersion systems, can be exploited.   With this instrument many studies have been made on metals, crystals, distorted surfaces, indenta-tions, etc.   The basic optics of the Linnik design (several variants by others have also been described) is shown in Fig. 16.4.   *Two high-power microscope objectives, as near identical as possible* are set up as shown. L_1 views the object O and L_2 views a polished

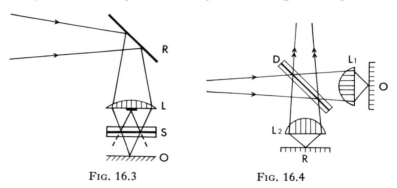

FIG. 16.3   FIG. 16.4

plate R chosen to have comparable reflectivity with O, if that be possible.   D is the divider, a half mirror between glass plates. It will be seen that we have here a direct Michelson system, but now the lenses can approach as near as needed to O and R for full oil immersion to operate.

There is a considerable technical difficulty in manufacture of this very useful instrument.   Both lenses L_1 and L_2 are in the interference paths and must therefore have very near identical glass thickness.   If this is not the case, then even when O and R are both flat planes, the differential glass thickness leads to ap-pearance or circular rings, making it extremely difficult to judge true flatness.   In practice, the lens-makers try out by trial and error a large number of so-called "identical" lenses (possibly hundreds) until an optically matched pair is found. Both oil films must also be identical when oil immersion lenses are used.

Transmission Systems for Transparent Objects

Large numbers of microscopic objects are transparent and are viewed by transmitted light, as a consequence of which special optical problems arise in interferometry. Simple direct interference between a transmitted and a double reflected beam is not practicable because of low visibility considerations. The problems encountered have been solved in several different ways, and as a consequence there are now available a number of quite

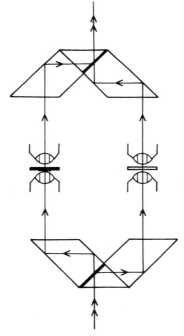

FIG. 16.5

different interference *transmission-light* microscopes of varying complexity, each of which can give useful information about small transparent objects. One can roughly classify these instruments into (i) those using *two* microscopes, (ii) those using *one* modified

microscope, (iii) interference *objectives*, (iv) *differential* interferometers. Many examples of each type are now commercially
available, but only representative instruments will be described.

Leitz make an instrument based on the interferometer of Mach-
Zehnder, already previously described. The arrangement is
shown (much simplified) in Fig. 16.5. This is a double microscope and the same difficulties as to careful lens matching exist
here as with the Linnik opaque-object instrument described
earlier.

A beam of light enters a beam splitter from below made up as a
double rhomb, the thick line being a 50% reflector. The two
separated beams go through two identical microscopes (the two
condensers must be matched as well as the two objectives). In each
beam is a glass slide, one holding the object. Oil immersion can
be used, and here a further difficulty is met with in that both oil
drops should be of identical thickness. Practically this is so difficult to achieve that instead, variable thickness compensators are
introduced in each path to enable optical path equality on both
sides to be secured. Such equality is necessary when white-light
fringes are required. These compensators can also be tilted and
this enables control over interference fringe dispersion. After
passage through the microscopes the beams are recombined at a
double rhomb similar to the first.

This arrangement is simply a Mach-Zehnder interferometer
wherein path differences are introduced by the refractive index and
thickness variations of the biological object on the slide. Suppose,
for instance, the object has a lenticular nuclear region, then
circular "Newton's rings" fringes will appear in this region. Or
again, since path differences can be balanced, fringes can be
obtained with white light. Then very slight variations in thickness or refractivity can cause marked *changes in the colour tint*
of the *transmitted* coloured fringes. There is, consequently, a
very great enhancement of contrast. For many years reasonably
sensitive colour contrast effects have been created in biological
objects by histochemical staining. But such techniques, sensitive
as they are, almost invariably kill the objects. On the contrary the
enhanced colour contrasts of interferometry can be obtained on
living objects, which is obviously a considerable experimental
advantage.

An additional cost with this instrument is that any change of objective lens for viewing requires insertion of an identical matched lens in the other path, hence if a set of objectives is required, each must be identically paired. The same is true for condenser lenses. As a consequence such an instrument with a comprehensive set of lenses is quite expensive.

Jamin Interferometer Type Microscopes

The earliest practical transmission interference microscope was that described in 1893 by Sirks. It is a simple Jamin interferometer mounted before the microscope, as shown in Fig. 16.6. Parallel light, from below, meets a glass block A_1. Most of the light goes through as beam 1, but some is reflected at the glass surface. Then, after meeting a small area of silver, it again

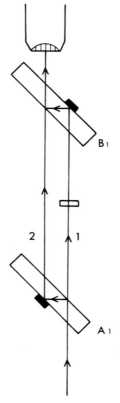

FIG. 16.6

reflects to form beam 2.   The light beams go through a glass slide one only passing through the object and one nearby.   Thus a path difference is introduced.   The block B_1 identical with A_1 brings the beams, now with the same intensity, into interference and fringes are then seen by the microscope.   These fringes give any microtopographical *thickness variations* in the object.

Control of fringe dispersion was achieved by introducing a small angle biprism into the beams 1 and 2.

Polarized light beam splitting

The Sirks-Jamin system necessarily required reasonably thick glass blocks to achieve sufficient sideways beam separation.   The microscope objective is therefore an appreciable distance from the object, consequently only low lens powers can be exploited.   An advance was made by Lebedeff (1930) by the introduction of beam splitting with doubly refracting crystals, *which need not be thick* and which are set perpendicularly to the microscope axis, instead of at 45°, as in the Sirks system.

When light strikes (at normal incidence) a calcite plate, suitably cut, it finds in the crystal its two privileged directions and splits therefore into two beams which are plane polarized at right-angles to each other.   The one beam (ordinary) passes straight through, the other (extraordinary), of same intensity, deflects and emerges displaced sideways from the ordinary beam.   The refractive indices are different for the two beams.   Lebedeff's instrument uses this in accordance with the arrangement shown in Fig. 16.7.

Light from below is plane polarized by A (nowadays this is a sheet of Polaroid material).   The emergent light has its polarization plane such that on entering the thin calcite plate B, this plane bisects the privileged directions in the calcite.   The incident beam consequently splits into two sideways separated beams, of equal intensity, but polarized mutually perpendicularly.   These beams pass respectively through the slide and the slide plus object and in due course are brought into interference through exploitation of the properties of the half-wave plate H, which is able to reverse the planes of vibration.

In Fig. 16.7 because the half-wave plate turns the vibration planes through 90°, when the light meets the second calcite crystal, C, the beams now converge and meet.   A second polarizer is set to select resolved components from the two beams and brings them together to produce interference fringes.

Looking more closely into Fig. 16.7 shows that the Lebedeff type of instrument leads to a doubling of the image. For although only an *ordinary* beam going through the object is shown there, it is clear from Fig. 16.8 that some distance to the right is a region (rays are shown dotted) in which on the contrary an *extraordinary* beam also goes through the object. Thus interference patterns

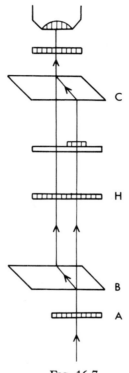

F ɪ ɢ. 16.7

are created by images 1 and 2. Fortunately, not only are these two images displaced *laterally*, but because of the differences of refractive indices, they appear in *different focus in line of sight*. According to whether the birefringent plate is thick or thin, the two images are completely separated or overlap.

If it is desired to examine a *small* part of a *big* object, then it can happen that the two rays, ordinary and extraordinary, are not sufficiently separated, so that one goes through one part of the object and the other through a neighbouring region of the object.

In this case only *differences* between the two regions are effective in producing fringes.

Several commercial instruments are now available based on the Lebedeff system.   In that of Zeiss, for example, special attention is devoted to production of all glass components free from strain, otherwise the null positions in polarized light are influenced. An extensive range of objectives is available.   The beam-splitting plate is thin and cemented directly on the front of the objective, reducing working distance.   It can also be rotated, with the lens. The lower beam-splitter can be tilted, thus altering the colour shown by an object.   Various compensators can be injected into

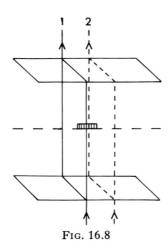

FIG. 16.8

that path which has no object, to enable a range of object thicknesses to be examined.   The manufacturers state that thickness changes *of only 25Å* in an object, when viewed in white-light fringes, can be measured by a setting of the compensator.

Differential Interferometers

In the polarization beam-splitting instruments just discussed, the beam-splitter (often a plate of calcite) has thickness such that the two plane-polarized beams which emerge are sufficiently sideways separated so that, with a small object, only one beam goes through the object, the other missing it.   A different situation exists if, by using a *very thin* calcite plate, the beam separation is so slight that both beams then go through the object which is under study.

Even if the two polarized beams do go through different regions of practically the same metrical thickness, there can still be an appreciable *path difference* because of the possibility of different refractive indices in the object for the two beams. It so happens, however, that of more importance than path differences are the *gradients* of path differences. It is not the change of thickness between adjacent regions that now matters, but the *rate of change*,

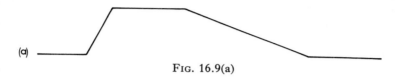

(a)

FIG. 16.9(a)

and because of this an interferometric system which deliberately uses only very slight sideways shearing of beams is called a "differential" interferometer. The characteristic optical properties of such a system are illustrated in Fig. 16.9.

Let us imagine an object which has a structure which converts a plane wave into the shape shown at (a). If a beam-splitter produces only a slight displacement, because there is also retardation in line of sight, the superposition of the two beams is as shown at

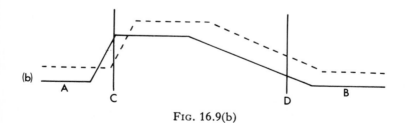

(b)

A   C   D   B

FIG. 16.9(b)

(b). If observation is with *white light*, then any small path differences can introduce *colour* changes. In the regions A and B the path differences are all the same, so no colour changes appear here. However, because of the slopes, the path differences at C and D are greater than those at A and B. The steeper the slope of the wave-front the bigger is the path change. Thus the bigger the gradient (i.e. rate of change) of the wave-front the more the colour change observed.

This differential principle can be applied in a relatively cheap and simple fashion in that it is possible to produce differential interference *eyepieces* which replace the ordinary occulars of a microscope.   More than one type is available, that of Françon being illustrated in Fig. 16.10.

An object O on a slide is illuminated from a slit aperture, with condenser lens L and viewed with the high-power microscope objective M.   The microscope *eyepiece* is replaced with the Françon interference eyepiece, as shown.   A is a lens followed by a Polaroid sheet P and after this is a beam-splitter C which is an arrangement called a Savart plate.   This plate consists of two identical thin pieces of birefringent quartz crystal, cut at 45° to the optical axes and set with polarization directions at right-angles to each other.   The plane-polarized beam entering the quartz splits into ordinary and extraordinary beams, *with only slight sideways displacement.*

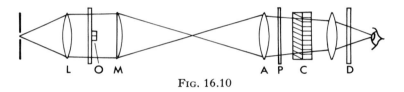

L   O   M   A P   C   D
FIG. 16.10

The second half of the Savart plate reverses the polarizations. By tilting this plate the path difference introduced is under control.   The two displaced beams are united by an analyser D, a sheet of Polaroid, crossed to that in front of the Savart plate. Thus a fringe system is created.

Other devices use Wollaston prisms instead of a Savart plate. This also produces beam displacement similar to that of a calcite plate.   As a rule the displacement chosen is selected such that the duplicated images are so near as not to be resolved as a pair by the microscope lens.   In a material of refractivity μ and of thickness which varies from t, it is not μt, which the fringes reveal but $d/dx(\mu t)$, hence the fringe patterns are quite different to any already met with.   Thus, whilst a spherical object, with ordinary μt fringes leads to Newton's rings, the differential interferometer shows for such an object equally spaced straight-line fringes running parallel to a diameter.   The numerical interpretation of differential fringes can involve integration methods.   One must be careful in making physical interpretations from these gradient fringes, since the physical microtopography is often not immediately self-evident as is the case with ordinary height contour fringes.

Differential eyepieces are cheap, robust, free from vibration problems and can be used both with opaque or transparent objects. They can produce striking pseudo-three-dimensional patterns on certain objects. Thus on viewing a curved platelet, like a blood corpuscle, the curved edge on the one side creates a highlight, whilst the oppositely curved edge (opposite gradient) on the other side produces a shadow. This leads to a pseudo-three-dimensional image which is quite striking. Hollows and hillocks, even if only slight, are thrown into brilliant relief; this is helpful in biological studies.

CHAPTER 17

HOLOGRAPHY

Introduction

The principles of holography, essentially an applied interferometric technique, were first laid down in 1949 by Gabor with the aim of improving electron microscope performance using what Gabor called "reconstruction of wave-fronts". This concept remained dormant, as it seemed at the time that it could only come to fruition by employ of what was not then available, namely, an *intense* beam of *coherent* monochromatic light. A close approximation to coherence does exist on the wave-front diverging from a pinhole, a fact used, of course, in the classical Young's slits experiment. However, the tiny apertures imposed of necessity lead only to quite low light intensities.

The invention of the source of light now universally called the laser (the name is composed of initials of the phrase, "light amplification by stimulated emission of radiation") created a new source of high-intensity, parallel, monochromatic light which has also a *very high degree of coherence*. The concept of the laser was first laid down by Townes in 1954, but it was first made to work as a light source by Maiman in 1960. There are now numerous kinds of lasers, all sources of coherent light, but of these our main concern is with two types, namely (i) short pulse *ruby* lasers, and (ii) continuous wave *gas* lasers. The pulsed ruby laser emits a flash of extraordinary brilliance which may be packed into no more than a hundred millionth part of a second or less. The gas laser, which is more frequently used in holography, emits a much less intense, yet still bright, *continuous* source of coherent and highly monochromatic radiation. It was the advent of these coherent sources which permitted the development of holography by Leith and Upatnieks in 1961 and since then the subject has developed at an explosive rate. Holography (from the Greek *holos* meaning whole) is an optical system which exploits the "whole" of the information in a wave, both *the phase* as well as the intensity, and it is this which gives it its special characteristics.

The Laser

In an atom in its normal state the outer (valency) electrons are in their *lowest* levels.   These can be lifted up to *higher* levels (excited states) by collision impact, or by suitable irradiation (fluorescence). As a rule a raised electron falls back very quickly indeed to the normal state and in so doing a pulse of radiation is emitted.   In characteristic excitation of a gas by fluorescent irradiation, normally only a small fraction of the atoms in the gas have their electrons raised to some higher excited level.   Yet if an *extremely powerful* flash of irradiation light is used, then, just following the exciting light flash, a situation can be created (if the source be bright enough) wherein there are actually *more* atoms with lifted-up electrons than are left with electrons in the normal state. Such a brief-lived, unusual situation is called a "population inversion".   In this condition, after an extremely small time-interval, an electron does fall back, and the atom emits as usual a single light pulse (a photon).   This photon on its way out of the gas container will now pass large numbers of other excited atoms. It interacts with one such excited atom and induces it to radiate. In fact it triggers off a photon.   Theory shows that the original trigger photon and the one which it induces both go in the same direction and, furthermore, are not only of exactly the same wavelength, but are also very closely coherent.

Each of these two coherent light pulses (photons) on its way out now repeats the process.   So, because of the population inversion offering the ready facility, in an extremely short time through a veritable chain reaction, an avalanche of light pulses (photons) is ejected in the one direction.   In effect this is an emission of intense, parallel, monochromatic, coherent light.   If the whole gas system be enclosed between two typical parallel Fabry-Perot type reflecting mirrors, then the coherent light reflects in multiple repeated fashion through the gas.   This acts therefore to amplify the whole situation.   The velocity of light is such that between mirrors, say, 3 cm apart there can be as many as 100 cross-traverses in a mere 10^{-8} second.   This avalanching mechanism builds up to a great intensity and it is called *lasing*.   This lasing mechanism was first successfully brought into action with the ruby laser.   Ruby is a crystal of aluminium oxide with a small amount of chromium metal (0.05%) distributed finely in the crystal in an atomic form, virtually as a free gas of chromium atoms.   It is the chromium atoms which are those excited into laser action by irradiation. The ruby is made synthetically by melting the oxide, and must be very pure and in rod form (some centimetres in length). The ends of the rod can be polished plane parallel and coated with silver to Fabry-Perot thickness. The chromium in the ruby is raised into population inversion by a fluorescence mechanism, nowadays called "optical pumping". The ruby rod is surrounded by a helical quartz gas discharge tube containing xenon and this is again surrounded by a polished reflecting cylinder. A massive, short, highly concentrated, discharge from a large bank of big condensers is sent through the xenon tube and so bright is the flash, (it may only last a millionth of a second), that immediate population inversion results, followed at an interval which can be below 10^{-9} second by a brilliant very short flash of laser light. Although little more than a thousandth part of the input flash energy converts to laser emission light, the short duration of the latter leads to an extraordinary brilliance. Thus, for instance, the laser energy coming out from a square centimetre of ruby face may be at the rate of 20 kW. The emerging light is parallel and can be brought to a spot focus by a lens, over an area as small as 1/100th of a square millimetre. This focused intensity is then 2 000 000 kW. This is many thousands of times brighter than sunlight and, furthermore, is *all light of one wavelength*, hence compared with the intensity of this corresponding wavelength in the continuum of sunlight, laser light is many millions of times more intense.

Yet this is far from the end of the intensity position, for it can be still much more heightened by what is called "giant pulsing". Ordinarily the lasing emission mechanism starts to develop before the optical pumping flash has completed its work. This leads to spread of lasing over a finite (even if only a very small) time-interval, with consequent reduction of peak intensity. If an extremely high speed shutter can be interposed between the mirrors, the lasing action can be delayed, functioning only when the shutter is open, to permit of build up by reflections. Such a shutter forces through the avalanche building mechanism in a shorter time than normal. This operation is usually called "Q-spoiling" and if correctly arranged, shortens the time of lasing and consequently builds up the intensity (the *total* output is not increased, but the peak brilliance is greatly heightened). The factor gain can be × 10 000. The shutter system used is a

Kerr cell, an electronic light chopper which is in essence a container with nitrobenzene between two plates. A high frequency alternating potential on the plates causes rapid variation of birefringence and, as laser light is polarized, this acts as a high-speed shutter. Only one face of the ruby is silvered, a Fabry-Perot reflector being placed near the other face, with the Kerr cell in between. Q-spoiling leads to very short flashes of extremely great peak intensity.

Although short-duration and high-intensity lasers are indeed used in very many applications such as, for instance, distance ranging between earth and moon, for holography (our main interest here) the continuous gas laser, though far inferior in intensity is most frequently used. When a correct mixture of the gases helium and neon, for instance, is maintained in a tube between Fabry-Perot mirrors then because of collision interaction a mixture of 10 parts helium and one part neon can be induced by a continuous electric discharge into a population inversion situation. Because of multiple reflector end mirrors, a lasing mechanism can be induced. Such a system lases with light in the red region of the spectrum. It gives out monochromatic parallel red light of fair brilliance (nothing like pulsed ruby rods, of course) yet coherence is excellent and the equipment is not expensive. It is such sources which are mostly used in holography, which essentially exploits the *coherence* of the light. Argon lasers can give continuous beams of blue light with a power of 50 W, whilst CO_2 gas lasers can emit as much as a kilowatt of power in the infra-red region.

Principle of holography

The hologram is most frequently a photographic plate which receives and records the light wave scattered or diffracted by an object. Normally, a photographic plate records only the *intensity* in the wave front. In order to include the influence of *phase* variations as well, an arrangement is made to record *interference fringes* on the photographic plate. This is carried through by simultaneously bringing on to the photographic plate another coherent plane wave. There are many different ways whereby this can be achieved, one such being shown in Fig. 17.1.

Parallel *coherent* light from a continuous wave laser is sent from the direction B on to the object (any object) which in general both scatters and diffracts light on to a photographic plate C. From the same original coherent laser source, (attenuated in intensity to be comparable to the light arriving on the photographic plate from the object), by means of reflector R, there is sent a second beam A which also reaches the photoplate. The two coherent beams, i.e. that scattered from D and that reflected from R, produce local interference fringes on the plate C in the form of a highly complex pattern of *close packed very complicated twisting narrow fringes.* The photo plate is then developed in the normal fashion and this is then called the "hologram", having on it, according to the object, a complex of fringes whose spacing, shape and intensities are created by both the *intensities* and the *phases* of the light scattered and diffracted by all the complex structural features of the object. The more complicated the object the more bewilderingly complex is the fringe pattern forming the hologram. This fringe pattern

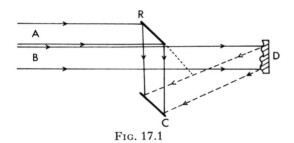

Fig. 17.1

bears *not the slightest resemblance* to the object itself, yet it carries *all* the optical information (both phase as well as intensity) given out by light scattered and diffracted by that object.

For the fringe positions are determined by the reflectivity, the diffraction angles and the phases. It will be noticed that the two interfering beams meet at a large angle, consequently the fringe dispersion is minute, i.e. fringes are very close packed. It is therefore necessary to use *very fine-grain* photographic plates if the fringe detail is to be faithfully recorded. This makes exposure times *long* since fine-grain photoplates are inherently insensitive.

If now, by reversing the situation, a hologram plate is illuminated at normal incidence with parallel monochromatic *coherent* light A as in Fig. 17.2, then an optical reconstruction is created. Most of the incident light passes through and is lost at B (hence one need for a *bright* coherent source). The close-packed complex of fringes on the hologram act as a very complex "diffraction grating", reversing the original situation and, as a consequence, two images form, one real at C, one virtual at D. The initial interfering beams of Fig. 17.1 meet at a large angle, so likewise the resulting close-line pattern of fringes on the hologram create the diffraction images also out at a wide angle, and thus they are very accessible and easy to see or record. Leaving aside any falsities introduced by imperfect photo-recording, etc, the diffraction image should be a true reconstitution of the object, since it is created from both phases and amplitudes of the light scattered by the object.

Of the two images created, real and virtual, the former can be photographed and the latter seen by eye. They are framed by the hologram plate. Because of inclusion of all the optical

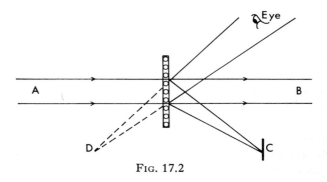

FIG. 17.2

information an astonishingly vivid three-dimensional reconstruction is seen. In consequence a viewing camera can be set *at any selected focal plane* to focus on selected different depth positions of the object, all lying latent within the *one* hologram. As an example suppose a thin cloud of droplets is produced in a cloud chamber. If this be illuminated to create a hologram, the reconstituted images of droplets can be studied at leisure at *selected different depths within the cloud chamber*. The one hologram contains within it the recording of *all* depth positions. Both transparent and opaque objects can give usable holograms and the hologram itself is observable both in transmission (Fig. 17.2) and also in reflection. The real image formed by a hologram has what is called a pseudoscopic character, that is, it appears with a reversed curvature. This is eliminated, if required, by making a second hologram from the original hologram, the double image reversal now restoring the correct curvature. Because the real reconstituted image is in the air, distant from the hologram plate, a camera or microscope can be brought within very close range and this is highly advantageous in many cases.

A curious situation arises when the real original object is brought into coincidence with the aerial holographic image, for thereby *low-order* (white-light) interference can now be produced to show the microtopography of the object.

Experimental Imperfections

Since *all* photographic reproduction suffers from grain imperfections, plate errors, intensity irregularities and non-linear blackening, there are inevitably losses in recording and this shows itself by imperfections in the final reconstituted image. Furthermore, the finite size of the hologram plate itself affects the observable aperture and this limits resolution of fine detail. A further difficulty, resulting from a peculiar property of highly coherent light, is the appearance of what is called "speckle pattern". If a typical non-specular scattering object (most objects are diffracting and non-specular) is looked at with coherent light, the high coherence allows of interference between light waves scattered, even diffusely, from *separate* close-by regions. So, consequently, a rough object shows a random pattern of small bright speckles of light, which seriously confuse resolution of fine detail in the object. This same speckle also reappears in a reconstituted hologram and can be very confusing at times.

Another problem in holography is stability. It will be seen from Fig. 17.1 that the hologram is simply an interferogram, and this clearly depends critically upon rigid maintenance of the distance between object and recording photo-plate. Equally, the refractive index in the intervening space must also be held constant. Thus, any vibrations, or temperature or pressure changes, indeed even air draughts, must all be rigidly excluded. *Any* effective change of optical path between the object and the photo-plate *greater than only a tenth of a light wave* leads to blur in the reconstituted image. Then again, it is tacitly assumed that both the wavelength and the coherence of the laser source will not change during the reasonably long exposure needed to create the hologram. Since fine-grain plates are imposed, exposure times of the order of 10 minutes are often required and it so happens that many laser sources can drift over such a period of time.

Especially when big depths (say half a metre) are involved in the object then stringent coherency is an absolute essential. To render a gas laser very stable can considerably increase its cost. In the end, holography is a fairly expensive procedure when equipment cost and source replacements are taken into account, for laser sources have only limited lives.

Applications of holography

Resolution tests with good holograms show that the lines on an object 0·01 mm apart can be resolved. This is only one-twentieth of what a good optical microscope can resolve. The inability to utilize highly converging beams and the existence of speckle both contribute to the deficiency. Some investigators have reduced speckle by an averaging process through *vibrating* the hologram plate during an exposure, but this is a difficult procedure and not too satisfactory.

The *aerial* nature of the image permits of many unusual applications. If a deep hollow can at all be reached by light then the holograph image can be reached by a microscope, even if not accessible to the microscope on the real object. Then again, objects at any selected depth inside a transparent chamber can be photographed or microscopically examined. Composite repeat holograms of any particular surface can be superposed to produce notable effects of holographic interferometry. This can be done in a variety of ways. Thus, for instance, an exposure is made on a photo-plate to give the hologram of a complex surface. Without moving (or developing) the plate, the object surface is then slightly tilted and a second (tilted surface) hologram is now superposed on the first, the compound image on the plate then being processed. When there is reconstitution from this double hologram, the object surface is now seen covered over with a pattern of interference fringes, showing the surface microstructure. The advantages of such systems of holographic interferometry will now be discussed.

Holographic Interferometry

Holographic interferometry has an enormous advantage over ordinary two-beam classical interferometry in that fringes of good definition can be given by quite *rough non-specular objects*, such as roughly machined engineering components, which would fail entirely to respond to ordinary interferometric examination through being far too rough to give specular reflection. If the surface to be examined is indeed very rough and irregular then holographic coherence requirements impose that such a surface can only be matched with regions slightly displaced from themselves. In other words, the condition which existed in the differential interferometer (see Chapter 16) now exists, too, and as a consequence the difficulties in interpretation associated with seeing the real, physical meaning of differential fringes can arise in holographic interferometry also. Nevertheless, with holographic interferometry matt, rough, unpolished, crude surfaces can give surprisingly good microtopographic interference fringes.

Even for an irregular or complex surface, anything which slightly alters the surface, or alters its position, can produce clear holographic interference fringes. Elastic deformation of massive rough engineering components have yielded precision information by superposing on the one hologram plate the two wave fronts of reflected light, one before, the other after producing an elastic deformation. On reconstruction, the engineering component (however rough) is seen crossed by fringes giving the extent and form of the distortion suffered.

An alternative useful procedure involves the use of two *separate* holograms, the first being taken (and fully developed) with the unstressed object. A second hologram is made with the stressed object. The second hologram is now viewed *through* the first (unstressed) hologram and as a consequence an interference fringe pattern appears across the object image. Many vibrating objects (diaphragms, musical instruments, vibrating or rotating machinery) have been studied using this technique. For some rapidly rotating objects, the laser light can be stroboscopically chopped at the same speed as the rotor understudy. The hologram from the moving object is then viewed through a hologram previously made of the same object when it was at rest, and as a result a stroboscopically "frozen-in" interferogram results. As the stroboscopic frequency can be slightly changed, a slow run through of phase variations with time can be studied.

Plate VI (c) shows the holographic interference pattern given by the vibrating belly of a violin which is sounding. Such a wooden object, of such shape, would fail completely to yield results if ordinary optical interferometry had been attempted.

Up to now the discussion has considered *continuous* wave laser sources. If a *pulsed* laser source is used, so short is the time of the flash that even a fast-moving object is, relative to this, virtually at rest. Pulse times of 10^{-7} second are not difficult to obtain, indeed 10^{-9} second can be reached. If during this time, the object does not move more than a tenth of a light wave, a well-defined hologram can be secured. This principle has been used in unexpected fashion. Even those rapid reactions of chemicals both in gases and in solutions, which can cause rapid changes in refractive index, can still be followed and recorded by the fast laser pulses available. Clearly, many fast changes of different kinds lend themselves to analysis with pulsed lasers.

Some Further Properties of Holograms

It has already been indicated that images located at different depths can be reconstituted. In addition to this, by viewing off-axis, images and interferograms can be seen in different lines of sight. Since more than one image can be recorded in succession, there remains for the future the possibility of *holography in colour*. It will be necesary to record a set of holograms (on one plate) from three lasers of different (effectively primary) colours. Reconstruction with the same three, coloured laser sources, should create an image in its true colours, even if the holograph plate itself consists of fringe patterns which are *purely black and white*.

Holographic interferometry offers advantages in such fields as aerodynamic interferometry. To begin with the simpler Michelson set-up is possible instead of the usual more complex Mach-Zehnder interferometer. The choice of the Mach-Zehnder system for aerodynamics (as already explained) is due to the fringe localization condition. In the classical Michelson system fringes and object are in different localization planes. Not so in holography, where the plane selected for the fringes can be coincident with the object plane. A related method used in the National Bureau of Standards U.S.A., illustrated in Fig. 17.3, is simply a Michelson system. Light from the gas laser divides at a beam-splitter, part going to the reference mirror and part *through* the experimental chamber to another mirror. The beams re-unite on the holographic plate, as shown.

First a hologram is made of the system in the static unstressed position. Then the wind tunnel is set into operation, leading to expected refractive index changes in the light path. A second (altered) hologram is then superposed on the first. When this double combination hologram is reconstructed, then the required fringe pattern is seen, in this case, for the double pass of modified light through the gas chamber (there and back).

The advantages of this holographic system are twofold. First, there is the question of fringe localization, which here, as in all holograms, is entirely under control; second, because holographic interferograms *do not require optical flats*, the whole optical path (mirrors, lenses, windows) can be of quite ordinary (non-special) quality. For the same defects reappear in both holograms, identically which are being superposed. Therefore they do not contribute to those differences which lead to the fringe formation.

An additional advantage lies in the fact that the first hologram

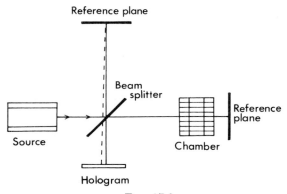

Fig. 17.3

recorded on the plate is from a surface which is nearly plane therefore the final fringe pattern seen is a match from the distorted path with a near plane wave so that the fringes, in this case, are not differential fringes, but are of the type normally encountered in non-holographic interferometry. This lends itself to easy interpretation.

Holograms without photographic plates

Materials other than photographic plates have been exploited for the recording of holograms. High resolution (grainless) holograms have been recorded thus. A glass plate is coated with gelatine containing a solution of potassium dichromate. It is found that when this plate is irradiated with light the gelatine hardens where the light falls. If this irradiated plate is then immersed in water a differential swelling results.   If this water is later abstracted with a dehydrant, an etched record of the incident illumination pattern is left behind.   This then can be used as a hologram.   It is a *phase* grating, not an *amplitude* grating.   Very fine detail (several thousand lines per millimetre) can be faithfully recorded by such a plate.   However, this kind of hologram suffers distortions from any subsequent atmospheric moisture absorption effects.   It remains yet to be developed fully.

A kind of hologram which may have technical potential has been called the "erasable" hologram, and it could find employ as an information store for computers.   One such erasable system employs what is called a "photochrome", a material which changes its absorption to light when it is illuminated.   With this a holographic image can be formed when light of suitable wavelength is used.   This same material can be rebleached with light of another wavelength, so that an erasing action can be carried out at will and the holograph recording device be reused indefinitely.

Another erasable type uses a thin transparent film of lithium niobate as the sensitive material.   Where this is illuminated, because of its photo-electric properties, electrons are locally excited and this causes local change in refractivity.   The optical path length changes locally and thus a hologram (also in phase instead of in amplitude) is created.   Illumination by ultra-violet light erases the image in this case.

Many other devices have been proposed, some far removed from ordinary photography.   In one system the light falls on to a thin film of magnetic material which has the character of changing its magnetic properties when illuminated, because it has a strong magneto-optical constant.   The changes in magnetization alter the magnetic domains in the film and it is this domain pattern which is the hologram.   In this case a magnetic field can erase the pattern very simply.

There is no doubt that many new techniques will emerge in the near future for a great deal of experimental work is in progress in holography, both as regards development of new holographic techniques and as regards applications to unexpected fields.

CHAPTER 18

SOME INTERFERENCE SPECTROSCOPES: ECHELON GRATING AND LUMMER PLATE

The Transmission Echelon Grating

The resolving power of a line grating is nm, where n is the total number of lines cut on the grating and m is the order of interference. It is found in practice that the cutting diamond with which gratings are made tends to fail when n approaches 100 000 lines and very few gratings can be made with more lines than this. The intensity in a normal grating falls off rapidly with order, and thus it is unusual to be able to go even up to $m = 4$. The intensity difficulty can be helped by a technique called "blazing". It can be so arranged to incline the cutting diamond such that a large amount of the incident light can be "blazed" into one selected order (often the second). Even so a resolving power of some 400 000 is about the attainable limit and this is only secured at the expense of much loss in light.

Michelson (1898) realized that a resolving power of say 400 000 can alternatively be secured by a grating of only 40 elements. ($n = 40$) if the order of interference m has the high value 10 000, and yet at the same time there results a great increase in intensity. This Michelson achieved by a simple but extremely ingenious device called the echelon grating, which consists of a series of stepped parallel plates of as near as possible identical thickness.

The stepped (echelon) arrangement is shown in Fig. 18.1. If a parallel beam of light is directed on to the face AB, the light beams emerging from successive steps are retarded behind each other by a constant equal to the optical path difference between the path in air and through a step. With a step of thickness of the order of a centimetre, the number of waves in the retardation path is very large (some 10 000, perhaps), i.e. m is very large. The number of beams n is equal to the number of steps. It should be as large as possible. In practice, in the transmission echelon, fifty-five plates is the maximum number that can profitably be used.   Beyond that point absorption of light in the long glass path, and losses by reflections at the many interfaces, begin to lead to too great a loss in light.   The mechanical difficulties in construction for n greater than 40 are considerable.

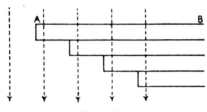

FIG. 18.1

The *transmission* echelon is costly to build, is sensitive to temperature fluctuations, and is in many respects inferior to the Fabry-Perot interferometer.   Yet its modification by Williams into the *reflection* echelon is an important development.   Although

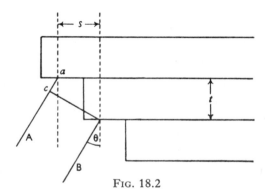

FIG. 18.2

the transmission echelon has largely dropped out of use, it was employed to some extent in the past and a reasonable number of such instruments are in existence.   It has not the flexibility of the Fabry-Perot interferometer, and, if constructed out of glass, is of no value for ultra-violet investigations.

Theory of the Transmission Echelon

Referring to Fig. 18.2, let s and t be the width and height of a single step.   Parallel incident light strikes the first surface normally.   Consider a beam diffracted through the angle θ.   The path difference between the two rays A and B from corresponding points is for reinforcement

$$\mu t - ac = m\lambda$$

where μ refers to the glass.

Now

$$ac = t.\cos\theta - s.\sin\theta$$

therefore

$$m\lambda = \mu t - t.\cos\theta + s.\sin\theta$$
$$= (\mu - 1)t + s\theta$$

as θ is small.

Thus

$$\theta = (m\lambda/s) - (\mu - 1)t/s$$

To get the dispersion $d\theta/d\lambda$, differentiate, giving

$$d\theta/d\lambda = (m/s) - (t/s)(d\mu/d\lambda)$$

as approximately

$$m = (\mu - 1)t/\lambda$$

($s\theta$ is small in comparison with $(\mu - 1)t$) then

$$\lambda.d\theta/d\lambda = (t/s)\{(\mu - 1) - \lambda.d\mu/d\lambda\}$$

The dispersion can be written as $d\theta/d\lambda = tb/\lambda s$, where

$$b = \{(\mu - 1) - \lambda.d\mu/d\lambda\}$$

If $d\theta_s$ is the smallest wavelength interval that can be resolved, it follows that the smallest resolvable wave-number interval, $d\nu$, (the resolving limit) is given by

$$d\nu = (s\nu/tb)d\theta_s$$

Since in this $d\theta_s$ is the angular increment necessary for resolution, then, according to the Rayleigh criterion for a rectangular aperture,

$$d\theta = \lambda/ns$$

for the quantity ns is the total width of the grating aperture A, and in accordance with theory the limiting angle of resolution is λ/A.

Thus we have:

$$d\nu = 1/ntb$$

The resolving limit $d\nu$ is therefore inversely proportional to the number of steps n, the thickness of the plates t, and the quantity b. Michelson takes for flint glass b as roughly unity. The second term in b is smaller than the first, and approximately, $d\nu = \nu/nm$, as in the line grating.

The distance between adjacent orders is obtained thus:

$$\Delta\theta = (\lambda/s)\Delta m$$

and as $\Delta m = 1$ for successive orders, $\Delta\theta = \lambda/s$ in which $\Delta\theta$ is the angle between successive orders. By equating $d\theta = \Delta\theta$ the wave-number range $\Delta\nu$ is obtained, namely:

$$\Delta\nu = (s\nu/tb)(\lambda/s) = 1/tb$$

Since, approximately, $b = m\lambda/t$ then

$$\Delta\nu \approx \nu/m$$

The larger the value of t the lower is the resolving limit, but then the smaller is the available range. A compromise must be adopted. The value b for flint glass varies from about 0·57 at 6000 Å to 0·73 at 2000 Å. Taking 0·66 as a rough average value, and comparing this with the Fabry-Perot interferometer, in which $\angle\nu = 1/2t$, it is seen that a Fabry-Perot interferometer, with gap of the same thickness as the transmission echelon plates, has a dispersion range about three times smaller than that of the echelon. Thus, for plates 1 cm thick, the range between orders is some 1·5 cm⁻¹.

The Intensity Distribution Within the Fringes

Since virtually there is diffraction from an aperture, all the light available can be concentrated into one or two orders. This leads to a great increase in intensity relative to the corresponding line grating. The intensity distribution of the light diffracted by a rectangular aperture is given by the expression:

$$I = A^2 \frac{\sin^2(\pi s\theta/\lambda)}{(\pi s\theta/\lambda)^2}$$

The intensity distribution has the form shown in Fig. 18.3. It is zero for $\theta = \pm\lambda/s$. All the light is practically included between the two points distant $\pm\lambda/s$ from the centre. The distance between successive spectra is λ/s. If approximately normal light is viewed, then there will be a strong central maximum at $\theta = 0$, and as the next orders fall at λ/s ($\theta = \pi$) they practically vanish. This is the single order position. (The black line in Fig. 18.3.)

By slightly tilting the echelon this central image moves away towards the point $-\lambda/s$ and a second image appears from the point $+\lambda/s$. At a suitable tilt two orders of equal intensity can be made visible. These are distant λ/s apart. This constitutes the double order position (see Fig. 18.4).

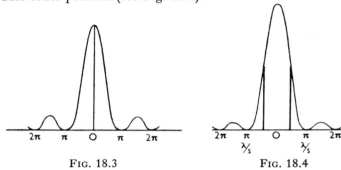

FIG. 18.3   FIG. 18.4

The greater the number of plates, the more are the inter-reflection losses and also the more accurately must the plates be worked. Michelson has shown that for n plates the retardations must differ by less than $1/n$th of a wave. Since the limit of optical accuracy is about one-eightieth of a wave, intensity considerations being disregarded, as many as eighty plates can in theory be used, but as yet the largest instrument made contains forty plates.

The Transmission Echelon in Practice

A typical instrument may have twenty-nine plates (thirty beams) have a thickness $t = 7$ mm and a step width $s = 1.5$ mm. The value of s cannot be reduced too far for then the aperture of the system is diminished, with consequent loss of light. Angular separation between orders being λ/s this becomes 3.3×10^{-4} radian for 5000 Å. Thus with a focal length of a projecting lens of as much as 3 metres the separation between orders is only 1 mm. This small dispersion is a serious drawback and also necessitates the use of very fine slits, and slow fine-grain photographic plates.

A convenient method of use is to place the instrument in the parallel beam of a spectroscope, between the lens of the collimator and the prism. The collimator must be moved to accommodate the interferometer. The echelon steps are placed to rise up as in a normal staircase. The spectroscope slit is opened fairly wide, but it is crossed with a fine *horizontal* slit.   Thus in the spectroscope the various lines are separated, and each shows the interference pattern for the echelon, but with the interferometric dispersion in the up-down direction in the field of view (on a photographic plate).

The dispersion is fixed and there can at times be grave difficulty in allocation or orders.   Indeed it is necessary to have two instruments of different dispersion to settle order allocations. The instruments are costly to make.   They are highly temperature-sensitive, have small dispersion and require a precision slit. They are subject to the defect of exhibiting ghost images, often arising through strains in clamping, supporting, or temperature differences.   These instruments are rarely used nowadays for serious researches.   Their typical effective use is for the measurement of hyperfine structure of spectrum lines, Zeeman effects, line widths, etc.   The latter involves careful computation, to allow for the fringe shape imposed on the instrument by diffraction.

The Reflection Echelon

An important development was the reflection echelon due to Williams (1926).   There are technical difficulties in making an echelon grating to operate in reflection, but these were overcome by Williams who used plates of fused silica, optically contacted together by "wringing".   The echelon is built up and the front

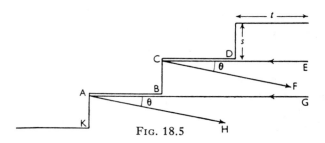

Fig. 18.5

face coated with a reflecting metal such as aluminium.   The plates must be worked to an exceptionally high standard of optical perfection and this makes the cost formidable.   The theory of the diffraction mechanism is similar to that of the transmission echelon, but simplified by the absence of a dispersing medium.   Consider Fig. 18.5.   A parallel beam of rays, denoted by EC, GA, falls normally on to the steps of a reflecting echelon face. The metallic coating only affects the *intensity* of the reflected light and is not involved in the interference mechanism. The beam diffracted at an angle θ is CF, AH. Let the distance AB = CD = t be the step thickness and AK = CB = s the step height. For θ small the path difference for successive rays is $\mu(2t-s\theta)$. For reinforcement this equals $m\lambda$, and considering $\mu = 1$ for air, we have $2t-s\theta = m\lambda$. The dispersion (neglecting the effect of the air) is obtained by differentiating, giving

$$d\theta/d\lambda = m/s = 2t/\lambda s$$

since m is sufficiently closely equal to $2t/\lambda$. If this is compared with the transmission instrument, for which

$$d\theta/d\lambda = bt/\lambda s$$

it is seen that the dispersion of the reflecting instrument is $2/b$ times as great, i.e. 3 to 4 times as great.

The angle between successive orders is, as in the former case, $\Delta\theta = \lambda/s$.

The resolving limit is $d\nu = 1/2tn$, in which n is the number of plates. It is therefore some three to four times better than that of the transmission instrument with the same thickness of plate t.

The Reflection Echelon in Practice

The reflection echelon is a valuable spectroscopic instrument. Its most striking characteristics are its range and its uniformity in performance. It is the only interferometer that can be used from the far infra-red to the very extreme vacuum ultra-violet. If introduced into a vacuum spectrograph it can in theory be used right down in the vacuum region. The limit of resolution, according to the simple theory, is independent of the wavelength. This is in actual practice not quite true. Apart from its use for the detection and measurement of hyperfine structures, Williams has shown that it can be used for the measurement of absolute wavelengths. In fact this would appear to be its most important use in the vacuum region. Being of fixed dispersion, it has all the attendant difficulties about allocation of orders, etc., characteristic of the transmission echelon. The number of high-resolution problems which can be solved with a single instrument alone is limited. In the transmission instrument a line is brought from the single- to the double-order position, and vice versa, by slightly tilting, so that the oncoming beam deviates from normal incidence. With the reflection echelon this cannot be done, for tilting throws the reflected beam out of the field of view. The difficulty has been overcome by enclosing the echelon in a gas-tight chamber. By making a small alteration in the gas pressure, the length of the optical path is changed and a line can readily be moved from single- to double-order position or vice versa. The gas chamber is not affected in its mode of operation by any temperature fluctuations, for as long as a given mass of gas is enclosed in a sensibly constant volume, the refractive index does not change with alteration in temperature.

A number of methods for using the reflection echelon have been described. One efficient method due to Jackson is shown in Fig. 18.6. The echelon is used as the back reflecting mirror of a Littrow spectrograph. An accurately made cross slit S is mounted

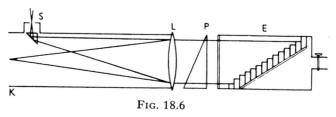

FIG. 18.6

horizontally on the slit of the quartz spectrograph. This is auto-matically therefore at the focus of the collimating lens L (quartz). A parallel beam falls upon the quartz prism P which has a re-fracting angle of 30°. Normally, in the Hilger type of E.I. quartz instrument, the back of this prism is silvered for the usual Littrow mounting. In this case the silvering is removed and parallel beams of light fall on to the reflecting echelon E. The light is back-reflected and fringes brought to a focus on the photographic plate K by the lens L which thus serves a double purpose.

The absence of additional optical components leads to high economy in light. There is a large saving in space and it is relatively easy to lag the straight 3-metre double air path for the purpose of temperature control.

The Lummer Plate Interferometer

Devised by Lummer (1901) and subsequently improved by Gehrcke (1903) the Lummer plate interferometer has had a con-siderable vogue but is much less used now than formerly.

As in the case of the Fabry-Perot instrument, the interference effects take place within a plane parallel plate, with the difference that, whereas in the former instrument the light is incident approximately in the normal direction, in the Lummer plate interferometer the light is incident at almost grazing angles. *An air Fabry-Perot interferometer employed with approximately grazing incidence has not the properties of a Lummer plate interferometer,* these being in effect a result of the dispersion of the glass or quartz. In both instruments, Fabry-Perot and Lummer plate, the high resolution obtainable is attributable to the summation of multiple-beams reflected at the plane parallel surfaces, the intensities of successive beams diminishing slowly because of an effective high reflecting coefficient. In the Fabry-Perot interferometer the summation is carried to infinity, but in the Lummer plate instrument a strictly limited number of beams is involved. There exists another important fundamental distinction. In the former instrument the high reflecting coefficient is obtained by the employment of a suitable metallic film, in the latter use is

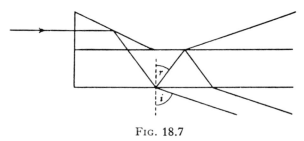

FIG. 18.7

made of the fact that near to total reflection the reflecting coefficient of any transparent substance is, in accordance with the well-known Fresnel expressions, close to unity, and is largely independent of wavelength. A beam of light striking a plane parallel plate near to grazing incidence suffers internal reflections and, on either side of the plate, beams emerge with successive constant retardation and reduction in intensity. If collected by a lens, fringes are formed at the lens focus, one set from either side of the plate. As in the Fabry-Perot interferometer these are fringes of equal inclination and require an extended source for a number of fringes to be seen.

The instrument is much more efficient if the light is introduced into the plate by a totally reflecting prism as shown in Fig. 18.7.

Then within the plate reflections occur near total reflection and high reflectivity results.

Theory of the Instrument

The instrument is used with a beam of parallel light incident nearly normal on to the face of the prism. With a plate of thickness t and refractive index μ, then as in the case of the Fabry-Perot parallel plate $n\lambda = 2\mu t . \cos r$ where r is the refracted angle in the plate.

Since
$$\sin i = \mu \sin r$$

$$2t\sqrt{(\mu^2 - \sin^2 i)} = n\lambda$$
$$4t^2(\mu^2 - \sin^2 i) = n^2\lambda^2$$

Differentiating gives:
$$\delta i = \frac{-n\lambda^2}{2t^2 \sin 2i} . \delta n$$

The fringes are of the equal inclination type and a change in angle, $\varDelta i$, corresponding to a change of a single order ($\delta n = 1$) is given by:
$$\varDelta i = \frac{-n\lambda^2}{2t^2 . \sin 2i} = \frac{-\lambda\sqrt{(\mu^2 - \sin^2 i)}}{t . \sin 2i}$$

Approximately, then:
$$\varDelta i = \frac{-\lambda\sqrt{(\mu^2 - 1)}}{t . \sin 2i}$$

i.e. $\varDelta i$ is proportional to $1/t$.

The dispersion, which is the rate of change of i with λ, is given by:
$$n^2\lambda = 2t^2\{2\mu(\partial\mu/\partial\lambda) - \sin 2i . \partial i/\partial\lambda\}$$
$$\therefore \frac{\partial i}{\partial\lambda} = \frac{4t^2\mu(\partial\mu/\partial\lambda) - n^2\lambda}{2t^2 . \sin 2i}$$

i.e.
$$\frac{\partial i}{\partial\lambda} = \frac{2\lambda\mu(\partial\mu/\partial\lambda) - 2(\mu^2 - \sin^2 i)}{\lambda . \sin 2i}$$

From this it will be seen that the dispersion is a function of the angle of incidence and the optical constants but does not involve the dimensions.

The wavelength range between successive orders is given by equating δi with Δi which gives $\Delta\lambda$ the order separation as:

$$\Delta\lambda = \frac{n\lambda^2}{n^2\lambda - 4t^2\mu \,.\, \partial\mu/\partial\lambda}$$

or alternatively in terms of wave numbers

$$\Delta\nu = \frac{n}{(n^2/\nu) - 4t^2\mu \,.\, \partial\mu/\partial\lambda}$$

Since $\sin^2 i$ is almost unity for the angles employed, then, closely enough,

$$\Delta\nu = \frac{1}{2t} \,.\, \frac{\sqrt{(\mu^2-1)}}{\mu^2-1-\mu\lambda \,.\, \partial\mu/\partial\lambda}$$

Owing to the excellent performance of the silvered Fabry-Perot interferometer, and because of the advantages of using a doubly refractive medium, only crystalline-quartz Lummer plate interferometers offer decided advantages. Crystalline quartz is doubly refracting, the dispersion depending therefore upon whether the ordinary or the extraordinary beam is used.

The Resolving Limit of the Lummer Plate Interferometer

The width of the wave-front entering the collecting lens is $l \cos i$ where l is the length of the plate. If this is regarded as an aperture A, then the angle between the central maximum and the first diffraction minimum is $\partial i = \lambda/A$, where ∂i is the limiting angle that can be resolved according to the Rayleigh criterion.

Since

$$\partial i = \frac{2\lambda\mu(\partial\mu/\partial\lambda) - 2(\mu^2 - \sin^2 i)}{\lambda \,.\, \sin 2i} \partial\lambda$$

equating $\partial i = \lambda/l \cos i$ gives $d\lambda$, the smallest wavelength separation that can be resolved. Converting into wave numbers, we have for the resolving limit:

$$d\nu = \frac{\sin i}{l(\mu^2 - \sin^2 i - \lambda\mu \,.\, \partial\mu/\partial\lambda)}$$

For grazing emergence $\sin^2 i = 1$, and as the quantity $\lambda\mu \,.\, \partial\mu/\partial\lambda$ is small we have:

$$d\nu_G = 1/l(\mu^2-1)$$

where $d\nu_G$ refers only to grazing emergence.

This simple expression shows that the resolving limit in this special case is independent of the wavelength (except in so far as μ depends on λ).

The expression derived for the resolving limit holds only in the regions close to grazing emergence, all the beams being considered to have about the same intensity. For larger angles this is not true. However, the smaller the emergence angle the higher is the effective reflecting coefficient, and as successive beams are then more equal in intensity, the sharper are the fringes, and consequently the lower the resolving limit. The Fresnel reflecting coefficients for the extraordinary rays are appreciably greater than those for the ordinary beams.

At an emergence of $1°$ the effective reflectivity is $0·94$ for the extraordinary ray but only $0·87$ for the ordinary ray. The superiority of the reflecting coefficient for the extraordinary beam is such that this is always used. With a Lummer plate interferometer constructed of glass the light used is polarized with a Nicol prism; with a quartz instrument the light is either first polarized, or alternatively selective use is made of the natural double refraction. The quartz plate is best cut with its optic axis on the surface, perpendicular to the long edge of the plate. Because the beams in the series stop at the end of the plate the resolution is inferior to that of a corresponding Fabry-Perot interferometer and, furthermore, weak secondary maxima form between the main fringes.

The Lummer Plate in Practice

The technical demands on optical perfection with a Lummer plate undoubtedly exceed those for all other optical requirements. Not only must the surfaces be plane to at least $\lambda/80$, they should be as parallel as possible (although a slight wedge angle is not too serious), and in particular the material must be as highly uniform as possible, otherwise ghost images can arise. Strains through supports must be avoided, and temperature control should be to within at least $0·04°C$. Early instruments had evil reputations for ghosts through inattention to one or more of these possible defects.

As with the transmission echelon the instrument can be placed in the parallel beam of a spectroscope, between collimating lens and prism. Since the fringes are those of equal inclination, no cross slit is needed and a wide open spectroscope slit is desirable (also as in the case of a Fabry-Perot interferometer) when crossed with a spectroscope.

The ordinary and extraordinary images can be separated either with a polarizing prism, or even merely by exploiting the double refraction of the quartz material of the plate itself. This leads to two separated images, which can be isolated by cutting down the *length* of the spectrometer slit.

As with the echelons a serious drawback is the lack of flexibility imposed by the fixed dispersion. The principal attraction of the instrument is its high performance in wavelength regions where the reflectivity of silver is too poor to permit the use of a silvered Fabry-Perot interferometer. Its performance in the ultra-violet can also be superior to that of an aluminized Fabry-Perot instrument of corresponding dispersion.

Techniques have in the past been devised for the detection of ghosts, but since the best available instruments are now ghost free (as revealed by the examination of single lines) these techniques have to-day little significance. Instruments are available with resolving powers of some 600 000. With these admirable experimental work can be carried out. Such a resolving power means that 0·01 Å can be resolved at $\lambda 6000$. Lummer plates have been used in a variety of investigations. Primarily they are used for the study of hyperfine structure in line spectra, but in addition many investigations into Zeeman effect have been made. One notable case was the evaluation of the strength of the magnetic field on the sun from the very small Zeeman effect produced on Fraunhofer lines in the sun's spectrum.

BIBLIOGRAPHY

THE number of original papers in interferometry is formidable and runs into thousands, published in a variety of languages. Instead of quoting extensive lists of these, a number of books are given with some indication of the scope and extent of the individual references in these books.

(1) *Handbuch der Physikalischen Optik.* E. Gehrcke (two volumes, in German; Barth, Leipzig, 1927). The long article by Feussner and Janicki on interferometry is a most learned and detailed account, containing over a thousand references.

(2) *Les Applications des Interferences Lumineuses.* Ch. Fabry (in French; Revue d'Optique, Paris, 1923). A monograph written with great clarity, concentrating somewhat on the applications of the Fabry-Perot interferometer. It contains 140 references.

(3) *Die Anwendung der Interferenzen.* E. Gehrcke (in German; Vieweg, Braunschweig, 1906). A monograph surveying earlier work and more detailed with respect to the Lummer plate interferometer. There are 150 references.

(4) *Studies in Optics.* A. A. Michelson (Chicago Univ. Press, Illinois, 1927). A masterly exposition of Michelson's own vitally important contributions. No references are given.

(5) *Applications of Interferometry.* W. E. Williams (Methuen, London, 1928). A valuable monograph. It contains 100 references; particularly useful on instrumentation.

(6) *Modern Interferometers.* C. Candler (Hilger and Watts, London, 1951). Somewhat specialized on instrumentation, containing 830 references; meant rather for the specialist.

(7) *High-resolution Spectroscopy.* S. Tolansky (Methuen, London, 1942). Deals in detail with the interferometers used for spectroscopic work where high-resolving powers are needed. In particular the Fabry-Perot, Echelon and Lummer Plate interferometers are treated.

(8) *Multiple-beam Interferometry of Surfaces and Films.* S. Tolansky (Clarendon Press, Oxford, 1948). This gives the theory and practice up to its date of publication of the use of multiple-beam interferometry, especially for the study of surface topography and measurements on thin films.

(9) *Surface Microtopography*. S. Tolansky. Longmans, 1960. More extended applications of multiple beam interferometry.

(10) *Interference Microscopy for the Biologist*. S. Tolansky. C. C. Thomas, 1968. Development and applications of interference microscopes to transparent objects, mainly those of interest in biological sciences.

(11) *Multiple Beam Interference Microscopy of Metals*. S. Tolansky. Academic Press, 1970. Special developments of multiple beam interference methods for opaque objects, metals in particular.

INDEX